BORATES: CHEMICAL, PHARMACEUTICAL AND PHARMACOLOGICAL ASPECTS

MATERIALS SCIENCE AND TECHNOLOGIES

Additional books in this series can be found on Nova's website under the Series tab.

Additional E-books in this series can be found on Nova's website under the E-books tab.

MATERIALS SCIENCE AND TECHNOLOGIES

BORATES: CHEMICAL, PHARMACEUTICAL AND PHARMACOLOGICAL ASPECTS

IQBAL AHMAD,
SOFIA AHMED,
MUHAMMAD ALI SHERAZ
AND
FAIYAZ H. M. VAID

Nova Science Publishers, Inc.
New York

For permission to use material from this book please contact us:
Telephone 631-231-7269; Fax 631-231-8175
Web Site: http://www.novapublishers.com

NOTICE TO THE READER

The Publisher has taken reasonable care in the preparation of this book, but makes no expressed or implied warranty of any kind and assumes no responsibility for any errors or omissions. No liability is assumed for incidental or consequential damages in connection with or arising out of information contained in this book. The Publisher shall not be liable for any special, consequential, or exemplary damages resulting, in whole or in part, from the readers' use of, or reliance upon, this material. Any parts of this book based on government reports are so indicated and copyright is claimed for those parts to the extent applicable to compilations of such works.

Independent verification should be sought for any data, advice or recommendations contained in this book. In addition, no responsibility is assumed by the publisher for any injury and/or damage to persons or property arising from any methods, products, instructions, ideas or otherwise contained in this publication.

This publication is designed to provide accurate and authoritative information with regard to the subject matter covered herein. It is sold with the clear understanding that the Publisher is not engaged in rendering legal or any other professional services. If legal or any other expert assistance is required, the services of a competent person should be sought. FROM A DECLARATION OF PARTICIPANTS JOINTLY ADOPTED BY A COMMITTEE OF THE AMERICAN BAR ASSOCIATION AND A COMMITTEE OF PUBLISHERS.

Additional color graphics may be available in the e-book version of this book.

LIBRARY OF CONGRESS CATALOGING-IN-PUBLICATION DATA

Borates : chemical, pharmaceutical, and pharmacological aspects / Iqbal Ahmad ... [et al.].
   p. cm.
Includes index.
ISBN 978-1-61209-573-8 (hardcover)
1. Borates. I. Ahmad, Iqbal.
QD181.B1B647 2011
546'.67124--dc22
                 2010054078

Published by Nova Science Publishers, Inc. † New York

CONTENTS

PREFACE

Borates are among the most widely used compounds in chemical and related fields. They have extensive applications in consumer products including pharmaceuticals, cosmetics, insecticides, preservatives, detergents, bleaching agents, enamels, etc. These compounds have important implications in the fields of chemistry, biology, medicine, pharmacy, pharmacology, toxicology, agriculture and environmental sciences. The major objective of this work is to provide a thorough background of the chemical, pharmaceutical and pharmacological aspects of borates and their industrial applications. It contains extensive information for academic and industrial research workers in their fields of interest and gives them a collective view of the importance and current state of knowledge in this vastly growing subject. It is also intended for graduate students engaged in higher studies and research in their respective fields.

Chapter 1 deals with an introduction to borates, borate compounds, boron toxicity, effects on plant growth, uses and literature on borates. Chapter 2 covers chemical aspects of borates including the analysis of hydroxy compounds based on borate complexation using chromatographic, spectrometric, electrochemical and electrophoretic methods, borate interactions, analysis of borate compounds, borate catalyzed / inhibited reactions, and synthesis of borate compounds. Chapter 3 presents an account of pharmaceutical aspects of borates such as the medicinally important compounds, clinical uses, effect on drug stability, dissolution kinetics, pharmacokinetics, and yeast growth. Chapter 4 deals with pharmacological aspects of borates and includes their toxicity, absorption, metabolism, oxidative stress, enzyme inhibition, bioactive compounds and analysis in biological fluids. Chapter 5 is concerned with the applications of individual borate compounds and as preservatives, fire retardants, insecticides, pharmaceuticals, cosmetics and borosilicate glass.

The authors are grateful to Prof. Dr. S. Fazal Hussain and Prof. Dr. M. Aminuddin of the Faculty of Pharmacy, Baqai Medical University, Karachi, for helpful suggestions on various aspects of this work. One of us (I.A.) would like to extend his great appreciation to his wife, Shamim, for her patience and understanding during the preparation of this monograph.

Iqbal Ahmad
Sofia Ahmed
Muhammad Ali Sheraz
Faiyaz H.M. Vaid

January 2011

Chapter 1

INTRODUCTION TO BORATES

Boron occurs naturally in the environment in soil, water and plants. It is present in sea water at a concentration of about 5 mg B/L and in fresh water up to 1 mg B/L. The World Health Organization (2004) defines a boron level of 0.3 mg/L as the non-observed effect level (NOEL) for drinking water. Higher boron levels in drinking water may be toxic to humans. It is an essential micronutrient for the growth of plants, however, it can be harmful to some plants in high concentrations. Humans obtain their daily requirements of boron, mostly from fruits and vegetables. The predominant boron species in aqueous solutions is present in the form of boric acid.

1.1. BORATE COMPOUNDS

Boron is a non-metal and exists only in the +3 oxidation state. It does not occur free in nature and is found in combination with oxygen as oxyacids including metaboric acid (HBO_2), boric acid or orthoboric acid (H_3BO_3) and tetraboric or pyroboric acid ($H_2B_4O_7$), and borates such as sodium borate or borax ($Na_2B_4O_7.10H_2O$), sodium perborate ($NaBO_3.4H_2O$), calcium borate ($CaB_4O_7.4H_2O$) and zinc borate ($2ZnO.3B_2O_3.3.5H_2O$). Borates are mainly the salts of tetraboric acid which on neutralization with base yields crystalline salts. Sodium borate is found in large quantities in California as a crystalline deposit. It is obtained by leaching the earth impregnated with borax using water and evaporating the solution to yield crystals of sodium borate. It is also obtained from naturally occurring calcium borate by treating with sodium sulfate. Boric acid is produced from sodium borate or other borates by reacting with hydrochloric acid or sulfuric acid.

1.2. PHARMACEUTICAL IMPORTANCE

Boric acid, sodium borate and sodium perborate are pharmaceutical necessities and are mild antiseptics (United States Pharmacopeia, 2007; British Pharmacopoeia, 2009). These compounds have several clinical applications as antibacterial agents, antifungal agents and chemopreventive agents and are used in the treatment of wounds, eye and ear infections. The borohydride salts are used in the manufacture of hormones. Borate buffers may have a catalytic or stabilizing effect on the degradation of pharmaceutical compounds. The pharmacokinetics of boric acid and sodium borate in humans and animals has been studied.

1.3. ANALYTICAL APPLICATIONS

Boric acid and borates are important complexing agents and have extensively been used for the separation, identification, determination and structural investigations of dihydroxy and polyhydroxy compounds. Boric acid gel and boron containing chiral stationary phases are used for the chromatographic separation and determination of primary amine-containing compounds, sugars and glycosides. A large number of natural products, drug substances and enantiomers have been separated and determined by capillary zone electrophoresis and micellar electrokinetic chromatography by borate complexation. Various analytical techniques have been used for the determination of boric acid and borates and their complexes in biological materials. Boric acid has been used as a modifier in the determination of various elements (Ahmad *et al.*, 2009, 2010a).

1.4. BORON TOXICITY

The industrial use of boric acid and sodium borate represents the major boron chemical exposure to humans and the environment. They are completely absorbed by the oral route of exposure. Boron exposure may be caused through:

1. Consumption of private, municipal or commercial sources of drinking water;
2. Dietary consumption of crops and other food stuffs; and
3. Inhalation of boron compounds during mining, manufacturing and other industrial processes.

Dietary exposure of 3.1 mg B/d is considered as the best estimate of exposure. Adults in the US may typically ingest from 3.5–7.5 mg B/d that may be toxic (Moore, 1997). The Environmental Protection Agency (EPA) health advisory committee recommends boron concentration in drinking water not to exceed 0.6 mg B/L (0.06 mM B) over a life time exposure. Boron via food, water and consumer products represents the greatest source of exposure to borates (Richold, 1998).

Anhydrous boric acid (B_2O_3) has low acute oral toxicity; LD_{50} in rats is 1970 to 2100 mg/kg of body weight. It has low acute dermal toxicity; LD_{50} in rabbits is greater than 2000 mg/kg of body weight. The inhalation of boric acid by rats at an exposure concentration of 77 mg m^{-3} does not indicate any acute toxic effects. Studies on boric acid feeding in rat, mouse and dog, at high doses, have demonstrated reproductive and developmental effects on fertility, fetus and skeleton (U.S. Borax, 2000; Ahmad et al., 2010b). The LD_{50} of sodium borate in rats is 5.66 g/kg of body weight (O'Neil, 2001). The acute oral mean lethal dose of boric acid in 1-day-old chickens has been found to be 2.95 ± 0.35 g/kg of body weight. This classifies boric acid as only slightly toxic to chickens (Sander et al., 1991). EPA findings have indicated potential developmental toxicity of borates in animals. A panel reviewing National Toxicology Program (NTP) reproductive toxicants have identified boric acid as high priority for occupational studies to determine safe versus adverse reproductive effects (Robbins et al., 2009).

Field and laboratory investigations have demonstrated that the presence of soluble mineral dust in air can be sensed by humans via chemesthesis. These dusts have been found to stimulate the mucous of the upper airways (Cain et al., 2004). About half a million Americans are exposed to boric acid in the industrial work place, where it is used in the manufacture of various materials. The presence of boric acid dust (10 mg m^{-3}) may give rise to certain symptoms leading to irritation (Cain et al., 2008).

A study of the risk assessment of boron in glass wool insulation has shown that the estimated boron intake from inhalation of glass wool fibers in occupational settings is insignificant and without any health risks (Jensen, 2009). Pollution caused by boric acid can be controlled by irreversibly fixing during adsorption and co-precipitation using synthetic allophone-like nanoparticles (Opiso et al., 2009), and on adsorption by hybrid gels (Liu et al., 2009) and activated alumina (Bouguerra et al., 2009). Dimethylamine borane, a reducing agent used in electric plating of semiconductors, is highly toxic to humans through any route of exposure and dermal absorption is the major route of neurotoxicity. It induces acute cortical and cerebellar injuries and delayed peripheral neuropathy (Tsan et al., 2005).

1.5. EFFECT ON PLANT GROWTH

There are widely different boron requirements and tolerances to plants, depending on the local conditions and the availability of other nutrients. The concentration of boron in plant leaves has been found to vary from 2 to 3875 ppm, and mild leave injury may occur at or below boron concentration resulting in the greatest growth. The tissues levels less than 15 or 20 ppm B indicate deficiency, while levels in excess of 200 ppm may produce symptoms of boron excess (Garrett, 1998).

Boron-deficient pumpkin plants exhibit reduced growth, and their tissues are brittle. The leaf cell walls of these plants contain less than one-half the amount of borate cross-linked rhamnogalacturonan II (RG-II) dimer than normal plants (Ishii et al., 2002). Excess boron (20 mM) as boric acid reduces polyphenol oxidase activities in embryos and endosperm of maize seed during germination and affects growth (Olcer and Kocacaliskan, 2007). It is estimated that up to 17% of the barley yield losses in Australia are caused by boron toxicity (Miwa et al., 2007). Niacin may be involved in the reduction of toxic effects of boron on carrot by regulating growth metabolism (Demiray and Dereboylu, 2006).

A major intrinsic protein NIP5;1 and a boric acid / borate exporter, BOR1, proteins of *Arabidopsis thaliana* are essential for efficient boron uptake and plant development under boron limitations (Takano et al., 2006, 2008). A study of BOR1 and NIP5;1 demonstrates the importance of selective endocytic trafficking in polar localization and degradation of plant nutrients transporters for radial transport and homeostasis of plant mineral nutrients (Takano et al., 2010). The enhanced expression of NIP5;1 results in improved root elongation under boron limiting conditions in the plant (Kato *et al.*, 2009). NIP6;1 is a boric acid channel for preferential transport of boron to growing shoot tissues in Arabidopsis (Tanaka *et al.*, 2008). The activity of specific membrane components can be influenced by boric acid regulating the functions of certain aquaporin isoforms and ATPase as possible components of the salinity tolerance mechanism in maize roots (Martinez-Ballesta *et al.*, 2008).

1.6. USES

Borate compounds have extensive applications in a large number of chemical and related industries. They are used as pharmaceuticals, buffering agents, bleaching agents, insecticides, herbicides, dyestuffs, wood preservatives, fire retardants, wax emulsifier and fuel and lubricating oil additive. They have applications in ceramics, detergents, electrolytic refining, electroplating, drug and polymer stabilization, fiber optics, fiber glass, leather tanning, textile finishing, glass and steel manufacture and other industrial processes. Potassium and sodium borohydrides are used in the synthesis of corticosteroids. Sodium and potassium borates have several applications in cosmetic preparations (Garrett, 1998; Ahmad *et al.*, 2010a).

1.7. LITERATURE ON BORATES

A number of handbooks (Will *et al.*, 1996; Garrett, 1998; Chung, 2010), monographs on history (Travis and Cocks, 1984; Woods, 1994; Flores, 2004), chemical and biological applications (Ali *et al.*, 2005), boron deficiency and cure (Shorrocks, 1989, 1997), geology and production (Barker and Lefond, 1985), mineralogy and geochemistry (Grew and Anovitz, 1996; Alonso, 1998; Flores, 2004), crystal chemistry (Leonyuk and Leonyuk, 1983), chemical analysis (Flores, 2004), industrial chemistry (Smith, 2002), borate glasses (Pye *et al.*, 1978; Lewis, 1989; Kimura and Miyazaki, 2008), and reviews on sources and chemistry (Farmer, 1982; Raymond and Butterwick, 1992; Woods, 1994), role in tooth whitening (Li, 1996), fire retardance of boric acid (Wang *et al.*, 2004), boric acid suppositories (Prutting and Cerveny, 1998), role in plant growth (Garrett, 1998; Maurel *et al.*, 2008), pharmacokinetics (Murray, 1998), toxicity and analytical applications (Culver *et al.*, 1994; Richold, 1998; Ahmad *et al.*, 2009, 2010a,b), pharmacological aspects

(Ahmad *et al.*, 2010b), boron-based wood preservative (Kartal, 2010), and uses (Rosenfelder, 1978; Garrett, 1998; Chung, 2010) have been published on boron and borate compounds.

REFERENCES

Ahmad, I., Sheraz, M.A., Ahmed, S., Vaid, F.H.M. (2009). Analytical applications of borates. *Mat. Sci. Res. J., 3*, 173-202.

Ahmad, I., Ahmed, S., Sheraz, M.A., Vaid, F.H.M. (2010a). Borate: toxicity, effect on drug stability and analytical applications. In M.P. Chung, Ed., *Handbook on Borates: Chemistry, Production and Applications*, New York, NY: Nova Science Publishers Inc., Chap. 2.

Ahmad, I., Ahmed, S., Sheraz, M.A., Iqbal, K., Vaid, F.H.M. (2010b). Pharmacological aspects of borates. *Int. J. Med. Biol. Front., 16*, 977-1004.

Ali, H.A., Dembistsky, V.M., Srebnik, M. (2005). *Studies in Inorganic Chemistry. 22 Contemporary Aspects of Boron: Chemical and Biological Applications*, Amsterdam: Elsevier.

Alonso, R.N. (1998). *Los boratos de la Puna*, Ed. Camara de Mineria de Salta, Argentina.

Barker, J.M., Lefond, S.T., Eds. (1985). *Borates: Economic Geology and Production*, New York: Society of Mining Engineers, AIME.

Bouguerra, W., Marzouk, I., Hamrouni, B. (2009). Equilibrium and kinetic studies of adsorption of boron on activated alumina. *Water Environ. Res., 81*, 2455-2459.

British Pharmacopoeia (2009). London: Her Majesty's Stationary Office, Electronic Version.

Cain, W.S., Jalowayski, A.A., Kleinman, M., Lee, N.S., Lee, B.R., Ahn, B.H. (2004). Sensory and associated reactions to mineral dusts: sodium borate, calcium oxide and calcium sulfate. *J. Occup. Environ. Hyg., 1*, 222-236.

Cain, W.S., Jalowayski, A.A., Schmidt, R., Kleinman, M., Magruder, K., Lee, K.C., Culver, B.D. (2008). Chemesthetic responses to airborne minerals dust: boric acid compared to alkaline materials. *Int. Arch. Occup. Environ. Health., 81*, 337-345.

Chung, M.P., Ed. (2010). *Handbook on Borates: Chemistry, Production and Applications*, New York, NY: Nova Science Publishers Inc.

Culver, B.D., Smith, R.G., Brotherton, R.J., Strong, P.L., Gray, T.M. (1994). Boron, In: G.D. Clayton, F.E. Clayton, Eds., *Patty's Industrial Hygiene and Toxicology*, 4th ed., Vol. II, Part F, New York, NY: John Wiley & Sons. Inc., Chap. 42.

Demiray, H., Dereboylu, A.E. (2006). The effects of excess boron with niacin on *Daucus carota* L. (carrot) root callus. *Acta Biol. Hung., 57*, 105-114.

Farmer, J. (1982). Structural chemistry in the borate industry. *Chem. Ind.*, No. 5.

Flores, H.R. (2004). *El Beneficio de los Boratos, Historia, Minerales, Yacimientos, Usos, Tratamiento, Refinacion, Propiedades, Contaminacion, Analisis Quimico*. Crisol Ediciones, Salta, Argentina.

Garrett, D.E. (1998). *Borates Handbook of Deposits, Processing, Properties and Uses*, Academic Press, San Diago, CA, Chap. 9.

Grew, E.S., Anovitz, L.M., Eds. (1998). *Boron Minerology, Petrology and Geochemistry*, Vol. 33, Mineralogical Society of America.

Ishii, T., Matsunaga, T., Iwai, H., Satoh, S., Taoshita, J. (2002). Germanium does not substitute for boron in cross-linking of rhamnogalacturonan II in pumpkin cell walls. *Plant Physiol., 130*, 1967-1973.

Jensen, A.A. (2009). Risk assessment of boron in glass wool insulation. *Environ. Sci. Pollut. Res. Int., 16*, 73-78.

Kartal, S.N. (2010). Boron-based wood preservatives and their uses, In: M.P. Chung, Ed., *Handbook on Borates: Chemistry, Production and Applications*, New York, NY: Nova Science Publishers Inc., Chap. 10.

Kato, Y., Miwa, K., Takano, J., Wada, M., Fujiwara, T. (2009). Highly boron deficiency-tolerant plants generated by enhanced expression of NIP5;1, a boric acid channel. *Plant Cell Physiol., 50*, 58-66.

Kimura, H., Miyazaki, A. (2008). Glass fabrication method as compared with crystal growth method on borate materials, In: J.C. Wolf, L. Lange, Eds., *Glass Materials Research Progress*, New York, NY: Nova Scientific Publishers, Inc., pp.187-209.

Leonyuk, N.I., Leonyuk, I.I. (1983). *Crystal Chemistry of Anhydrous Borates*, Moscow: MSU.

Lewis, M.H., Ed. (1989). *Glasses and Glass Ceramics*, London: Chapman & Hall.

Li, Y. (1996). Biological properties of peroxide-containing tooth whiteners. *Food Chem. Toxicol., 34*, 887-904.

Liu, H., Ye, X., Li, W., Dong, Y., Wu, Z. (2009). Comparison of boric acid adsorption by hybrid gels. *Desalination and Water Treatment, 2*, 185-194.

Martinez-Ballesta M.D.C., Bastias E., Zhu C., Schaffner A.R., Gonzalez-Moro, B., Gonzalez-Murua, C., Carvajal, M. (2008). Boric acid and salinity effects on maize roots. Response of aquaporins ZmPIP1 and

ZmPIP2, and plasma membrane H^+-ATPase, in relation to water and nutrient uptake. *Physiol. Plant, 132*, 479-490.

Maurel, C., Verdoucq, L., Luu, D.T., Santoni, V. (2008). Plant aquaporins: membrane channels with multiple integrated functions. *Ann. Rev. Plant Biol., 59*, 595-624.

Miwa, K., Takano, J., Omori, H., Seki, M., Shinozaki, K., Fujiwara, T. (2007). Plants tolerant of high boron levels. *Science, 318*, 1417.

Moore, J.A. (1997). An assessment of boric acid and borax using the IEHR evaluative process for assessing human developmental and reproductive toxicity of agents. Expert Scientific Committee. *Reprod. Toxicol., 11*, 123-160.

Murray, F.J. (1998). A comparative review of the pharmacokinetics of boric acid in rodents and humans. *Biol. Trace Elem. Res., 66*, 331-341.

O'Neil, M.J., Ed. (2001). *The Merck Index*, 13th ed., Rahway, NJ: Merck and Co., Electronic Version.

Olcer, H., Kocacaliskan, I. (2007). Excess boron reduces polyphenol oxidase activities in embryo and endosperm of maize seed during germination. *Z. Naturforsch. [C], 62*, 111-115.

Opiso, E., Sato, T., Yonedes, J. (2009). Adsorption and co-precipitation, behavior of arsenate, chromate, selenate and boron and with synthetic allopane-like materials. *J. Hazard Mater., 120*, 79-86.

Prutting, S.M., Cerveny, J.D. (1998). Boric acid vaginal suppositories: A brief review. *Infect. Dis. Obstet. Gynecol., 6*, 191-194.

Pye, L.D., Frechette, V.D., Kreic, N.J., Eds. (1978). *Borate Glasses: Structure, Properties, Applications*, Material Science Research, NY: Plenum Press.

Raymond, K., Butterwick, L. (1992). Perborate, In: O. Hutzinger, Ed., *Handbook of Environmental Chemistry*, Vol. 3, New York, NY: Springer-Verlag.

Richold, M. (1998). Boron exposure from consumer products. *Biol. Trace Elem. Res., 66*, 121-129.

Robbins, W.A., Xun, L., Jia, J., Kennedy, N., Elashoff, D.A, Ping, L. (2009). Chronic boron exposure and human semen parameters. *Reprod. Toxicol., 29*, 184-190.

Rosenfelder, W.J. (1978). The industrial uses of boron chemicals. *Chem. Ind., 12*, 413-416.

Sander, J.E., Dufour, L., Wyatt, R.D., Bush, P.B., Page, R.K. (1991). Acute toxicity of boric acid and boron tissue residues after chronic exposure in broiler chickens. *Avian Dis., 35*, 745-749.

Shorrocks, V.M. (1989). *Boron Deficiency, Its Prevention and Cure*, London: Borax Consolidated Ltd.

Shorrocks, V.M. (1997). The occurrence and correction of boron deficiency. *Plant and Soil, 193*, 121-148.

Smith, R. (2002). Boron oxide, boric acid and borates. *Ullmann's Encyclopedia of Industrial Chemistry*, Valencia, CA: Wiley-VCH.

Takano, J., Miwa, K., Fujiwara, T. (2008). Boron transport mechanisms: collaboration of channels and transporters. *Trends Plant Sci., 13*, 451-457.

Takano, J., Wada, M., Ludewig, U., Schaaf, G., von Wirén, N., Fujiwara, T. (2006). The arabidopsis major intrinsic protein NIP5;1 is essential for efficient boron uptake and plant development under boron limitation. *Plant Cell, 18*, 1498-1509.

Takano, J., Tanaka, M., Toyoda, A., Miwa, K., Kasai, K., Fuji, K., Onouchi, H., Naito, S., Fujiwara, T. (2010). Polar localization and degradation of Arabidopsis boron transporters through distinct trafficking pathways. *Proc. Natl. Acad. Sci., 107*, 5220-5225.

Tanaka, M., Wallace, I.S., Takano, J., Roberts, D.M., Fujiwara, T. (2008). NIP6;1 is a boric acid channel for preferential transport of boron to growing shoot tissues in Arabidopsis. *Plant Cell, 20*, 2860-2875.

Travis, N.J., Cocks, E.J. (1984). *The Tincal Trial, A History of Borax*, London: Harraps Ltd.

Tsan, Y.T., Peng, K.Y., Hung, D.Z., Hu, W.H., Yang, D.Y. (2005). Case report: The clinical toxicity of dimethylamine borane. *Environ. Health. Persp., 113*, 1784-1786.

U.S. Borax Inc. (2000). Anhydrous boric acid, Material Safety Data Sheet, Occupational Health & Product Safety Department, Valencia, CA.

United States Pharmacopeia 30 / National Formulary 25 (2007). Rockville, MD: United States Pharmacopeial Convention, Electronic Version.

Wang, Q., Li, J., Winandy, J.E. (2004). Chemical mechanism of fire retardance of boric acid on wood. *Wood Sci. Technol., 38*, 375-389.

World Health Organization (2004). *Guidelines for Drinking-Water Quality*, 3rd ed., World Health Organization, Geneva, Vol. 1, pp 282-283.

Will, R., Ishikawa, M., Riepl, J., Schneider, J., Willhalm, R. (1996). Boron minerals and chemicals, *Chemical Economics Handbook*, SRI International, Menlo Park, CA.

Woods, W.G. (1994). An introduction to boron: history, sources, uses and chemistry. *Environ. Health. Perspect., 102*, 5-11.

Chapter 2

CHEMICAL ASPECTS OF BORATES

Borates are among the most widely used compounds in chemical and related fields. They are involved in chemical and pharmaceutical analysis, chemical synthesis, chemical catalysis, chemical and metallurgical processes and many other fields. These aspects have been described in the following sections.

2.1. CHEMICAL ANALYSIS

Boric acid and borates are extensively used as complexing agents in the identification, determination and structural investigations of hydroxy compounds including diols, polyols, sugar alcohols, disaccharides, oligosaccharides, polysaccharides, glycosides, nucleotides and other compounds by the application of chromatographic, spectrometric, electrochemical, electrophoretic and thermal methods. Boric acid gels have been used for the separation, isolation and determination of drug substances containing diol grouping such as dopamine, epinephrine and methyldopa. Boric acid is also used as a modifier in the determination of several elements. The methods of analysis of boric acid / borates and their complexes are presented below:

2.1.1. Qualitative Analysis

The qualitative analysis of borates is carried out by performing the official (Pharmacopoeial) and non-official identification tests. These tests are described as follows:

a) An acidified methanolic solution of boric acid / borate, on ignition, burns with a green-bordered flame (British Pharmacopoeia, 2009; United States Pharmacopeia-National Formulary, 2007).

b) To an aqueous solution of sodium borate, phenolphthalein solution is added. A red color is produced which disappears on the addition of glycerol (British Pharmacopoeia, 2009).

c) To an acidified aqueous solution of borate, few drops each of iodine solution and polyvinyl alcohol solution are added. An intense blue color is produced (United States Pharmacopeia-National Formulary, 2007).

d) A method of qualitative analysis of boric acid in biological material has been developed to identify the cases of poisoning. The blood or urine is acidified with concentrated hydrochloric acid and a drop of the sample is placed on the curcuma (turmeric) paper. After drying at room temperature, a red stain appears if boric acid is present. The detection limit of boric acid is about 0.1 mg/mL (Yoshida *et al.*, 1989a).

e) A histochemical staining method has been devised for the detection of boric acid in the tissue. Frozen 12-14 micron sections of the tissue, fixed in anhydrous ethanol, are stained in an acidified curcumin solution. After washing in acetic acid, a red stain indicates the presence of boric acid. The color is developed on the formation of rosocyanin by the reaction of boric acid and the protonated curcumin (Yoshida *et al.*, 1991).

2.1.2. Quantitative Analysis

Boric acid is a weak acid (pKa 9.2) and its official assay is based on treating the substance with mannitol (British Pharmacopoeia, 2009) or glycerin (United States Pharmacopeia-National Formulary, 2007) in aqueous solution, in which it acts as a strong monoprotic acid and can be titrated with standard alkali to phenolphthalein end point. Sodium borate is determined by a similar method (British Pharmacopoeia, 2009) or by direct titration with hydrochloric acid (United State Pharmacopeia-National Formulary, 2007). Polyhydroxy compounds such as glycerol and mannitol form 1:1 or 1:2 molecular complexes between the hydrated borate ion and 1, 2- or 1,3- diols and facilitate the release of a hydronium ion into the solution.

$$2\ \begin{array}{c}>\text{C (OH)}\\ \\ >\text{C (OH)}\end{array}\ +\ H_3BO_3\ \longrightarrow\ \left[\begin{array}{c}>\text{C}-\text{O}\diagdown\quad\diagup\text{O}-\text{C}<\\ \qquad\qquad\text{B}\\ >\text{C}-\text{O}\diagup\quad\diagdown\text{O}-\text{C}<\end{array}\right]^{-}\ +\ H_3O^+\ +\ 2H_2O$$

The method may be applied to commercial boric acid after suitable modification (Mendham *et al.*, 2000).

2.1.3. Borate Buffers

The use of buffers in chemical and pharmaceutical analysis may be necessary to maintain pH and to achieve chemical stability and solubility of reactants. The borate buffer systems used and their pH ranges are: Feldman's (pH 7.0-8.2), Atkins and Pantin (pH 7.6-11.0), and Gifford's (pH 6.0-7.8). In each of these systems the buffer pair is the borate anion and boric acid ($B_4H_7^{2-}$ / H_3BO_3), however, in the latter two systems, the sodium borate is formed *in situ* to participate in the reaction. In the borate buffer mechanism (Block *et al.*, 1974), several compounds including boric acid, metaboric acid, sodium metaborate, and sodium tetraborate are involved. Boric acid is a weak acid (pKa 9.2) and is ionized by the reaction shown by Eq. (1) or alternatively by Eq. (2), to indicate the intermediate conjugate base having a tetrahedral arrangement of hydroxyl groups around the boron (Edwards *et al.*, 1955). The product of the latter reaction (Eq. (2) is expressed as a Lewis acid-base adduct.

$$H_3BO_3 + H_2O \rightleftharpoons H_3O^+ + H_2BO_3^- \tag{1}$$

$$B(OH)_3 + 2H_2O \rightleftharpoons H_3O^+ + B(OH)_4^- \tag{2}$$

The reaction of boric acid in the buffer solution with sodium hydroxide gives rise to a salt (Eq. (3)) which decomposes to metaborate ($NaBO_2$) according to Eq. (4). The metaborate in the presence of excess boric acid forms the tetraborate salt as given by Eq. (5):

$$B(OH)_3 + NaOH \longrightarrow Na^+ + B(OH)_4^- \qquad (3)$$

$$NaB(OH)_4 \longrightarrow NaBO_2 + 2H_2O \qquad (4)$$

$$2NaBO_2 + 2B(OH)_3 \longrightarrow Na_2B_4O_7 + 3H_2O \qquad (5)$$

The sodium borate reaction with acid in the buffer involves the formation of the slightly ionized tetraboric acid ($H_2B_4O_7$) from the highly ionized sodium tetraborate present (Eq. (6)). The tetraboric acid is then hydrolyzed to give metaboric acid and boric acid (Eq. 7).

$$2HCl + Na_2B_4O_7 \longrightarrow H_2B_4O_7 + 2NaCl \qquad (6)$$

$$H_2B_4O_7 + 3H_2O \rightleftharpoons 2HBO_2 + 2B(OH)_3 \qquad (7)$$

The borate buffers have been found to operate best in the pH range of 7 to 11, with greatest buffer capacity at about pH 9 (Block et al., 1974). Sodium borate is used as a pH standard for the calibration of glass electrode. A 0.01 M solution of $Na_2B_4O_7.10H_2O$ has a pH of 9.18 at 25°C.

Borate buffers have extensively been used as complexing agents, mobile phase in chromatography and as an electrophoretic medium. Several applications of borate buffers are reported in Section 3.1 and 3.4. Some further uses of borate buffers are described as follows:

An analytical method for the determination of reductive sulphur [S(IV) and S(–II)] in glass has been reported. Sulphur is separated by distillation as hydrogen sulphide, trapped in buffered boric acid-zinc acetate solution and determined after conversion to an ethylene blue dye (Rohr et al., 2004). Recombinant *Bacillus halmapalus* alpha-amylase has been crystallized using MES-HEPES-boric acid buffer at 303 K (Lyhne-lversen et al., 2006).

Lysine and valine-rich peptides undergo a beta-hairpin transition followed by intermolecular self-assembly into a fibrillar hydrogel network. The hydrogel undergoes a stiffening transition on cooling below the critical temperature only when boric acid is used to buffer the peptide solution (Ozbas et al., 2007). The production of o-diphenols by immobilized mushroom tyrosinase is carried out in borate buffer at pH 9.0 (Marin-Zamora et al., 2009).

In the in-capillary determination of creatinine with electrophoretically mediated microanalysis, an increase in the concentration and / or pH of the borate buffer (background electrolyte) results in the improvement of reaction efficiency (Mason *et al.*, 2009).

The triangle and tetrahedron optimization methods have been used for the selection of background electrolytes including borate buffer in capillary zone electrophoresis (Sun *et al.*, 2008). The performance of common capillary electrophoresis buffers has been tested for the pKa determination of several types of compounds (pyridine, amines and phenols) in the pH range 3.7 to 11.8 (Fuguet *et al.*, 2008).

A novel procedure for *in situ* synthesis of a complex chiral selector, di-n-butyl-tartrate-boric acid complex by the reaction of di-n-butyl-tartrate with boric acid in a running buffer has been reported and its application in the enantioseparation of beta-blockers and structurally related compounds by chiral microemulsion electrokinetic chromatography has been demonstrated (Hu *et al.*, 2009).

2.1.4. Analysis of Borate Complexes

The application of various techniques in the analysis of borate complexes (Sciarra and Monte Bove, 1962; Saxena and Verma, 1983; Mendham *et al.*, 2000; Ahmad *et al.*, 2009) is described in the following sections.

2.1.4.1. Chromatographic Techniques

a. Gel Chromatography
Boric acid gels have been used for the separation and isolation of drug substances. These gels are based on immobilized alkyl boronic acids and have a selective affinity for 1, 2- or 1, 3-diol grouping such as those present in dopamine and in sugars or glycosides. The analyte is loaded onto the gel in a buffer (pH 7.0) and the borate complex formed is then broken down by using a mildly acidic eluent such as dilute acetic acid (Watson, 2005). A new type of hydrogel, based on polyvinyl alcohol-borate system, has been developed. The rheological and solubilizing properties of the hydrogel can be modified by the addition of a cosolvent to water (Carretti *et al.*, 2009).

An improved method of gel chromatography has been described for the determination of norepinephrine in human plasma. The human plasma is applied to boric acid gel column without deproteinization and pH adjustment, and norepinephrine is eluted with 1.3 M acetic acid in methanol solution. The pentafluropropionic derivative of norepinephrine is analyzed by gas chromatography- mass spectrometry (Yoshida *et al.*, 1980). A liquid chromatographic method for the simultaneous determination of norepinephrine, epinephrine, dopamine and serotonin in human plasma has been reported. These analytes are purified on boric acid gel, separated on a reversed-phase C_{18} column and detected electrochemically at +600 mV (Imai *et al.*, 1988). A phenyl boric acid column has been used for the separation of serum beta-methyldigoxin, digoxigenin, and metabolites prior to their determination by enzyme immunoassay (Higashi *et al.*, 2003a). The method has been used in therapeutic drug monitoring and pharmacokinetic studies of beta-methyldigoxin in serum (Higashi *et al.*, 2003b). Boric acid forms complexes with 2- hydroxycarboxylic acids. This principle is used for the isolation of vanillylmandelic acid, *p*-hydroxymandelic acid, vanillyllactic acid, and *p*-hydroxyphenyllactic acid in urine using a phenylboronate gel column and n-butanol as the mobile phase (Higa and Kishimoto, 1986).

A new method based on the combination of exclusion and ion exchange chromatography for the separation of 2-acetamido-2-deoxymannopyranoside-containing saccharides using borate-saturated polyolic exclusion gels has been developed. NMR studies of the interactions of N-acetylhexosamines with borate confirmed the importance of a proper stereochemical arrangement of acetamido sugars for their interaction with borate anions (Petraskova *et al.*, 2006). The structure of a scleroglucan / borax hydrogel has been investigated for its suitability in drug release studies of model compounds such as theophylline, vitamin B_{12} and myoglobin. The molecular dynamic analysis can shed some light on the anomalous swelling of the hydrogel, suitable for drug release of the tested compounds (Palleschi *et al.*, 2006).

b. Ion Chromatography

A specific analytical method for the determination of borates in caviare fish has been developed using ion-exclusion chromatography. Isolation of the analyte as a 2:1 sorbitol-borate complex was accomplished using an anion exchange column with conductivity detection. Recoveries of borates from fortified samples varied from 94.5 to 101.1% (Carlson and Thompson, 1998). Chromatographic boron isotope separation at 5 and 17 MPa with diluted boric acid solution using a strongly basic ion exchange resin column has been achieved (Musashi *et al.*, 2008).

An accurate and sensitive method based on the combination of pyrohydrolysis-ion chromatography (PH-IC) has been proposed for the separation and determination of boron as borate, chloride and fluoride in nuclear fuel. An isocratic elution with a mobile phase of mannitol-sodium bicarbonate is used for the IC separation. The detection limit of boron as borate is 24 µg/L (Jeyakumar *et al.*, 2008).

c. Gas Chromatography

A specific method based on sodium tetraethyl borate ethylation of organotin compounds followed by gas chromatographic analysis with pulsed flame photometric detection has been developed. It has been applied to the determination of highly toxic organotin compounds in harbour sediment samples (Bravo *et al.*, 2004). The use of room temperature ionic liquid as matrix medium for the determination of residual solvents in pharmaceuticals by static headspace gas chromatography has been reported. The limits of detection of residual solvents are in the ppm level (Liu and Jiang, 2007).

d. High Performance Liquid Chromatography

Boromycin, a macrodiolide, exists as a hydrophobic complex formed from boric acid and a chiral polyhydroxy macrocyclic ligand. It is covalently bonded to silica gel through a urea linkage to an attached d-valine ester. As a chiral stationary phase for liquid chromatography, it shows pronounced enantioselectivity towards primary amine-containing racemates with 98% separation (Wang *et al.*, 2007). Boric acid has been used as a mobile phase additive in the high performance liquid chromatographic (HPLC) analysis of ribose, arabinose and ribulose mixtures obtained from bioisomerization processes using an Amiscex MPX-87K column. The sugars, on complexing with boric acid, become negatively charged and elute faster from the column by means of ion-exclusion and are separated by the difference in their complexation capacity (De Muynck *et al.*, 2006). An improved HPLC method has been used for the assay of urinary catecholamines including norepinephrine, epinephrine and dopamine. The catecholamines are extracted by ion-exchange chromatography and eluted with boric acid. After paired ion separation, quantitation is carried out by coulometric detection. The interference of anti-TB drugs in the assay is removed by treatment with alumina (Manickum, 2008). In the determination of urinary free catecholamines, the urine is purified on a column of immobilized boric acid, catecholamines are separated by ion-pair reversed phase HPLC and detected electrochemically. The detection limit of catecholamines is from 1 to 5 µg/L (Speck *et al.*, 1983).

Removal of borate from radiolabeled gangliosides is achieved by its complexation with mannitol. The resulting mannitoborate complex is separated by reversed phase HPLC and the labeled gangliosides are repurified by ion-exchange and silica gel chromatographies (Yohe, 1994).

The indirect determination of trichlorfon by HPLC, based on its catalytic effect on the oxidation of benzidine to 4-amino-4'-nitro biphenyl in the presence of sodium perborate, has been made. The oxidation product is detected at 365 nm (Zhu et al., 2007). The bioavailability of chondrosine in mouse blood plasma has been studied by HPLC using boric acid as an eluent (Kusano et al., 2007). The evaluation of titanium and titanium oxides as chemo-affinity sorbents for the selective enrichment of organic phosphates by HPLC using boric acid as the sample loading solution has been conducted (Sano and Nakamura, 2007). A comparative reversed phase HPLC method for rapid identification of glycopeptides and its application in off-line liquid chromatography-matrix assisted laser desorption/ionization-mass spectrometry (LC-MALDI-MS) analysis has been reported. It is used in the analysis of glycoproteins including ribonuclease B, AGg1, ovalbumin and asialo fetuin using borate buffer as eluent (Kanie et al., 2008). A reversed-phase isocratic HPLC method has been developed for the analysis of linezolid antimicrobial in pig pulmonary tissue. The tissue samples and controls are prepared in borate buffer (Guerrero et al., 2010).

The dissociation constants of commonly used buffering species including boric acid have been determined for several acetonitrile-water mixtures. These pKa values have been used to predict the pH variations of the most commonly used HPLC buffers when the composition of the acetonitrile-water mobile phase changes during the chromatographic process, such as ingredient elution (Subirats et al., 2009).

e. Thin-Layer Chromatography

A thin-layer chromatographic system using boric acid impregnated plates has been developed for the separation of a mixture of phospholipids including phosphatidylinositol, phosphatidylserine, and phosphatidylglycerol. The method has been applied to resolve the liver, kidney and platelet phospholipids of the rat into six major classes (Fine and Sprecher, 1982). Radiolabeled phosphorylated derivatives of phosphatidylinositol have been separated by thin-layer chromatography using a boric acid system (Villasuso et al., 2008).

A rapid and reliable high performance thin-layer chromatographic method has been developed for the identification of *Cimicifuga racemosa* and its most common adulterants in black cohosh by fingerprint profiles. It can be used for quality control of the raw material in a current Good Manufacturing Practices environment (Ankli et al., 2008).

2.1.4.2. Spectrometric Techniques

a. UV and Visible Spectrometry

Several methods have been reported for the UV and visible spectrometric determination of boric acid or other drugs by complexation with boric acid in foods, biological materials, pharmaceutical formulations, plant materials, eye lotions, enzymes and natural waters. Borates can be directly chelate extracted from foods with 2-ethyl-1,3-hexanediol in n-hexane-n-butyl acetate mixture (8:2, v/v), transferred into sodium hydroxide solution and acidified with hydrochloric acid. The solution is reacted with curcumin and the color is measured spectrometrically. The detection limit of boric acid is 1 ppm and the method has been applied to the determination of boric acid in frozen prawns, shrimp and jelly fish (Ogawa *et al.*, 1979). An improved curcumin method has been used for the determination of boric acid (0.01-0.5 mg/mL) in biological materials in the cases of poisoning. Boric acid is extracted with an acidified 2-ethyl-1,3-hexanediol solution in chloroform and color is developed with curcumin dissolved in a mixture of acetic and sulfuric acids. The excess of protonated curcumin is destroyed with ethanol and the rosocyanin formed is measured at 550 nm (Yoshida *et al.*, 1989b). Boric acid at μg or mg/mL level in plants and in natural waters can be dissolved in 1-6 M hydrochloric acid, extracted into a 2,2,4-trimethyl-1,3-pentanediol solution in chloroform, treated with a solution of carminic acid in sulfuric and glacial acetic acids (1:2, v/v) and the absorbance measured at 549 nm. The molar absorptivity of the borate complex is 2.58×10^4 L mole^{-1} cm^{-1} and the Beer's law is valid for 0.05–0.4 μg/mL boron range (Aznarez *et al.*, 1985).

An assay method has been developed for the selective determination of drugs containing a 1,2-diphenolic moiety by the measurement of difference in absorbance at 242 nm between two equimolar solutions of the drugs in pH 7 phosphate buffer, one of which contains 0.1 M boric acid to form an ester. The method has been applied to the assay of adrenaline, isoetharine, isoprenaline, levodopa and methyldopa in pharmaceutical formulations (Davidson, 1984). A simple and reliable sequential injection analysis (SIA) system has been reported for the determination of boron as boric acid in eye lotions. The method is based on the complexation of d-sorbitol with boric acid followed by acid-base reaction with methyl orange and the measurement of absorbance at 520 nm. The Beer's law is valid up to 12 mg/L. The system has a detection limit of 0.06 mg/L and is able to monitor boron at a frequency of 30 samples per hour with a RSD of less than 0.6% (Van Staden and Tsanwani, 2002).

A novel spectrometric method for the determination of aminophylline, on reaction with boric acid, in pharmaceutical and mixed serum samples has been developed. The Beer's law is obeyed in the concentration range of 0.2 to 200 μg/mL. The method has a RSD of 0.28% (Li et al., 2009a).

A spectrometric assay has been used to study the formation of a trypsin-borate-4-aminobutanol ternary complex. The assay using acetyl arganine p-nitroanilide as the chromogenic substrate in combination with ^{11}B and ^{1}H NMR indicates a cooperative bonding interaction in which the borate is esterified by the oxygen atoms of the 4-aminobutanol and trypsin residue Ser (195) (London and Gabel, 2002).

A diode-laser-based sensor has been developed to measure nitric oxide mole fractions in coal-combustion exhaust using sum-frequency mixing of a 395 nm external cavity diode laser and a 352 nm laser in a beta-barium-borate crystal (Anderson et al., 2007). A multicomponent spectrophotometric method has been used to determine riboflavin and photoproducts in photolysis reactions carried out in the presence of borate buffer at pH 8.0-10.5. Boric acid forms a 1:1 complex with riboflavin (Ahmad and Vaid, 2006; Ahmad et al., 2008, 2009, 2010).

The J-aggregation behavior of diprotonated tetrakis (4-sulfonatophenyl) porphyrin [$(H_2TPPS_4)^{2-}$] in aqueous solution in the presence of the hydrophilic ionic liquid 1-butyl-3-methylimidazolium tetrafluoroborate exhibits a peak at 490 nm to account for the formation of $(H_2TPPS_4)^{2-}$ J-aggregates. In addition the results also showed decreased fluorescence emission and intensified Raman scattering peaks (Wu et al., 2008a).

A spectrophotometric method has been developed for the determination of traces of phosphorous in zirconium based alloys. It involves selective fluoride complexation controlled by boric acid. The phosphomolybdate formed is extracted into n-butyl acetate, ion associated with crystal violet and the absorbance measured at 582 nm (Dhavile et al., 2008).

b. Infrared (IR) / Fourier Transform Infrared Spectrometry (FTIR)

The effect of boron chloride formation from zinc borate in burning PVC has been studied by infrared spectral analysis. It is found that when boron halide is produced, the B_2O_3 glass layer is destroyed and boron is volatilized which is unfavorable to flame retardancy (Yang et al., 1999).

The structures of calcium fructoborate (Wagner et al., 2008), ionic liquids (1-butyl-3-methylimidazolium iodide and 1-butyl-3-methylimidazolium tetrafluoroborate and their aqueous mixtures) (Jeon et al., 2008a) and 1-butyl-3-methylimidazolium tetrafluoroborate and water mixtures with varying concentrations (Jeon *et al.*, 2008b) have been studied by infrared and Raman spectroscopy. Changes in the spectral shape in the OH stretch vibration regions show that the addition of borate causes rapid breakdown of hydrogen-bonding network of water molecules.

FTIR spectrometry has been applied to study the effect of mannitol and sodium borate on the freezing of bovine serum albumin. FTIR analysis of the secondary structure of bovine serum albumin has shown an apparent protein structure-stabilizing effect of amorphous mannitol and sodium tetraborate combinations during the freeze-drying process. The mannitol and sodium tetraborate combination also protects lactate dehydrogenase from inactivation during freeze-drying (Izutsu *et al.*, 2004).

Attenuated total reflectance Fourier transform infrared (ATR-FTIR) spectroscopy has been employed to investigate mechanisms of boric acid and borate adsorption on hydrous ferric oxide surface. Infrared spectra from an adsorption isotherm of boric acid on hydrous ferric oxide show negative absorption peaks at 1625 cm^{-1} and at 1490 cm^{-1} due to the loss of water and carbonate, respectively, from the hydrous ferric oxide surface. The remaining peaks in the spectra occur in the region of trigonal B–O v_3 bands at 1395, 1330 and 1250 cm^{-1} and are the result of adsorbed boric acid on the surface (Peak *et al.*, 2003).

The structure of monohydrate potassium triborate, synthesized from potassium carbonate and boric acid, has been characterized by FT-IR and Raman spectroscopy. The analysis shows that three coordination B$_3$–O bond, four coordination B$_4$–O bond and hydroxyl and triborate anions exist in the formula of the molecule (K[B$_3$O$_4$(OH)$_2$]), which can lose a molecule of water (Zhang *et al.*, 2007a).

c. Raman Spectrometry

Raman spectra of supersaturated aqueous solution of MgO.B$_2$O$_3$-32% MgCl$_2$.H$_2$O during acidification and alkalization and dilution have been studied. The results indicate that the higher concentrations of cation are beneficial to the dissolution of boric acid and the polymerization of polyborate anion (Zhihong *et al.*, 2004). Raman spectra of dissolution and transformation of chloropinnoite in aqueous boric acid solution indicate that 2MgO.3B$_2$O$_3$.15 H$_2$O (Kurnakovite) is the phase transformation product (Xiaoping *et al.*, 2004). The structure of the calcium fructoborate has been determined by infrared and Raman spectrometry (Wagner *et al.*, 2008).

d. Nuclear Magnetic Resonance Spectrometry (NMR)

Dilute aqueous solutions of boric acid contain $B(OH)_3$ and $B(OH)_4^-$ species only. A rapid exchange on the NMR time scale produces a single peak in the ^{11}B NMR spectrum which moves up on an increase in pH. Concentrated solutions of boric acid contain polyborate ions as shown by the NMR and Raman spectra (Salentine, 1983).

Sugars with proper arrangement of hydroxyl groups have been found to form ester complexes with boric acid, for example, diacid disodium saccharate forms a 2:1 sugar acid: borate complex in aqueous solution. The ^{1}H and ^{13}C NMR spectra indicate the involvement of 3,4-diol groups of the tetrol fragment in coordinating with the boron atom (Van Duin *et al.*,1987). The interaction of boric acid and the polysaccharide guaran (major component of guar gum) has been studied by ^{11}B NMR spectrometry. A comparison with the ^{11}B NMR spectra of boric acid and phenylboronic acid complexes of 1,2-diols, 1,3-diols, monosacharides (mannose and galactose) and disaccharides (cellobiose and sucrose) indicate that the guaran polymer is cross linked via a borate complex of 1,2-diols both forming chelate 5-membered ring cycles. Based upon steric constraints it has been suggested that preferential cross-linking in the guaran polymer occurs via the 3,4-diols of the galactose side-chain (Bishop *et al.*, 2004). The interaction of hydroxypropyl guar with boric acid, lysozyme, and mucin has been studied to understand how hydroxypropyl guar interacts with tear film components. Borate binds to guar under the pH, temperature and ionic strength conditions similar to those found in the eye. The hydroxypropyl guar-borate complexes behave as anionic polyelectrolytes and thus interact with cationic lysozyme, a major tear film protein (Lu *et al.*, 2005).

The direct measurement of boric acid and borate adduction to NAD^+ and NADH by ^{11}B NMR spectrometry has been reported. The analysis shows that borate binds to both cis-2,3-ribose diols on NAD^+ forming borate monoesters, borate diesters, and diborate esters, whereas, only borate monoesters are formed with NADH. Boron shifts of borate monoesters and diesters with NAD^+ have been observed at 7.80 and 12.56 ppm at pH 7.0 to 9.0 (Kim *et al.*, 2003). Complexation of the ribose group of NAD^+ with boric acid is preferred over that of NADH on the basis of electrostatic interaction. This leads to the inhibition of the coenzyme system (Smith and Johnson, 1976).

The application of multiple quantum filtered (MQF) NMR to the identification and characterization of the binding of ligands containing quadrupolar nuclei to proteins is demonstrated. Multiple binding of boric acid and borate ion to ferri- and ferrocytochrome c has been detected by relaxation time measurements using MQF NMR. Borate ion has two different binding sites, one is in slow exchange and the other in fast exchange (Taler et al., 1999).

^{11}B NMR studies of the formation of ternary complexes of trypsin, borate and S1-binding alcohols reveal evidence of an additional binding interaction external to the enzyme active site. This binding interaction is a prototypical interaction of borate and boronate ligands with residues on the protein surface (Transue et al., 2006). The presence of two species in the catalyst prepared from B(OPh)₃ are revealed by ^{1}H NMR studies. One of the species can be formulated as a linear pyroborate on the basis of ^{11}B NMR spectrum (Zhang et al., 2008a). The determination of absolute configuration of alpha- and beta-hydroxyl acids by NMR via chiral BINOL borates has been conducted (Freire et al., 2008). Trisilyl-substituted vinyl cations have been isolated from hydrocarbon solutions as tetrakis(pentafluorophenyl)borates and identified by their characteristic ^{13}C and ^{29}Si NMR data (Klaer et al., 2009).

e. Mass Spectrometry (MS)

Matrix-assisted laser desorption / ionization Fourier transform mass spectrometry (MALDI / FTMS), operating in the negative ion mode, has been used to directly observe sugar alcohol borate complexes in plant extracts. In the mannitol borate complex, the most favorable configuration is with C-3 and C-4 of both mannitol residues complexed to the borate, thus allowing maximum interaction of the remaining hydroxyls with the borate centre (Penn et al., 1997). A method has been developed to characterize diols in boric acid complexes using negative ion electrospray mass spectrometry in combination with collision-induced dissociation tandem mass spectrometry (MS/MS). The electrospray mass spectra of acyclic vicinal diols indicate efficient complex formation. In the cyclic vicinal diols only the cis isomer produces an intense mixed complex, whose MS/MS spectrum is different from that of the trans form. The 1, 3- and 1, 4-diols are less prone to complex formation. The method may be used to find the structure of diols and polyols in aqueous solution (Acklov et al., 1999). Mass injection analysis with electrospray ionization mass spectrometry (EIMS) has been used to investigate borate-nucleotide complex formation. The borate complexation with nicotinamide nucleotide is significantly influenced by the charge on the nicotinamide group and the number of phosphate groups on the adenine ribose. Borate binding is decreased in the order of NAD^+, $NADP^+$ and NADPH (Kim et al., 2004). EIMS has shown that borate is bound to cis-2, 3-ribose diol and not to the hydroxyl groups on the phosphate backbone of NAD^+. The MS/MS spectrum indicates that in 1:1 NAD^+-borate monoester, the borate is bound to the adenine ribose (Kim *et al.*, 2003). A reflection time-of-flight mass spectrometer (RTOFMS) with a laser ablation ion source has been used to determine the isotopic ratio of $^{10}B/^{11}B$ present in boric acid solutions. The atom % ^{10}B values obtained were within ±1% of the actual ones (Manoravi *et al.*, 2005).

Fourier transform ion cyclotron resonance mass spectrometry (FTICR-MS) has been applied to identify boric acid / borate complexes with cafeic acid in various concentration ratios at pH 9.2. Experimental evidence of self-oligomerization of up to six borate units with cafeic acid, resulting in stable covalently bound polyborate-polyol complexes, is presented (Gaspar *et al.*, 2008).

The use of boric acid as preservative for urine samples containing trace elements leads to contamination in the analysis of trace elements by inductively coupled plasma-mass spectrometry (ICP-MS) (Bornhorst *et al.*, 2005). A method for the determination of volatile boric acid and total boron in boron nitride, boron carbide and boric acid using tungsten boat furnace (TBF) ICP-MS has been developed. It has been applied to the analysis of biological and steel samples (Kataoka *et al.*, 2008). The vaporization of boric acid has been studied by transpiration thermogravimetry and Knudsen effusion mass spectrometry (KEMS). The KEMS measurements performed for the first time on boric acid showed it as the principle vapor species with no meaningful information on water (g) though (Balasubramanian *et al.*, 2008).

The determination of total lead in lipstick by ICP-MS has been reported. Complete recovery of lead was achieved by adding hydrofluoric acid to the digestion mix, followed by treatment with excess boric acid to neutralize hydrofluoric acid and to dissolve insoluble fluorides. The average value of lead obtained for the lipstick was 1.07 μg/g (Hepp *et al.*, 2009).

Plant proteins of soy (glycinin, beta-conglycin) and pea (legumin, vicilin) have been identified by HPLC-MS. The method includes a pre-fractionation step to enrich plant proteins by using borate buffer (Luykx *et al.*, 2007). A mass spectrometric method has been developed for the identification of the carboxyl acid groups in analytes evaporated and ionized by electrospray ionization. It is based on gas-phase ion-molecule reactions of ammoniated and sodiated analyte molecules with trimethylborate and has been applied to the determination of non-steroidal anti-inflammatory drugs: ibuprofen, naproxen and ketoprofen (Habicht *et al.*, 2008).

A method based on headspace single drop microextraction (HS-SDME) in combination with gas chromatography (GC)-ICP-MS has been proposed for the analysis of butylin compounds in environmental and biological samples. Sodium tetraethylborate and sodium tetrahydroborate were used as the derivatizing agents for in situ derivatization of butyltins (Xiao *et al.*, 2008). A method for the trace analysis of methylmercury and mercury (II) in water samples has been developed. It involves stir bar sorptive extraction with in situ alkylation with sodium tetraethylborate and thermal desorption-gas chromatography-mass spectrometry (GC-MS). The limits of quantitation of methylmercury and mercury (II) are 20 and 10 ng/L mercury, respectively (Ito *et al.*, 2009).

Analytical methods based on ICP-MS have been developed for the determination of percutaneous absorption of ^{10}B in ^{10}B-enriched boric acid, borax, and disodium octaborate tetrahydrate in biological matrices (Wester *et al.*, 1998a, 1998b, 1998c). A method for the determination of boric acid in agar has been developed. After digestion with nitric acid the concentration of boron was measured by ICP-MS. The repeatability and reproducibility of the method were very good (Hamano-Nagaoka *et al.*, 2008).

Reactive desorption electrospray ionization (DESI) is a rapid and sensitive method for the direct detection of the hydrolysis product of phosphonate esters. The detection specificity is enhanced by implementation of a heterogeneous ion / molecule reaction using boric acid in the spray solvent. The reagent ion $H_2BO_3^-$ generated in the spray readily reacts with condensed phase alkyl methylphosphonic acids to form anionic adducts (Song and Cooks, 2007). Ultra-high-resolution mass spectrometry (HRMS) of polyborate species in aqueous solution has shown that these exist as clusters of polyborate anions (Gaspar and Schmitt-Kopplin, 2010). The formation of borate complexes of the peptide-derived Amadori products has been investigated by HRMS and tandem MS/MS (Kijewska *et al.*, 2009).

f. Fluorimetry

A fluorimetric method for the determination of rubomycin in biological materials in the concentration range of 0.04 to 20 µg/mL has been developed. After protein precipitation from the biological material with trichloroacetic acid, the antibiotic is determined by fluorescence emission of its borate complex in concentrated sulfuric acid using 575 nm as the excitation wavelength and 620 nm as the emission wavelength (Alykov *et al.*, 1976). Boric acid forms a complex with 2,2,4-trimethyl-1,3-pentanediol, which can be determined on the addition of a solution of carminic acid in a mixture of sulfuric and glacial acetic acids (1:2, v/v) by fluorescence emission measurement at 567 nm using an excitation wavelength of 547 nm. The fluorescence gives a linear response in the 8-120 ng/mL boron range (Aznarez et al., 1985).

The characterization of the dityrosine-borate complex has been carried out by fluorescence measurements. In the presence of excess borate/boric acid, the fluorescence emission maximum of the singly ionized dityrosine chromophore shifts from 407 nm to 374 nm. Fluorescence measurements performed as a function of pH and concentration are consistent with a 1:1 complex with maximal formation near pH 8 (Malencik and Anderson, 1991).

The fluorescence of putative chromophores in Skh-1 hairless mouse and calf skin acid-soluble type 1 collagens is due to tyrosine (λ_{ex} 275 nm, λ_{em} 300 nm), dopa (λ_{ex} 280 nm, λ_{em} 325 nm), tyrosine aggregate (λ_{ex} 280 nm, λ_{em} 360 nm), dityrosine (λ_{ex} 325 nm, λ_{em} 400 nm) and advanced glycation end product (λ_{ex} 370 nm, λ_{em} 450). These fluorophores can be markers of pathological conditions. Borate buffer quenches fluorescence at λ >400 nm from intact collagen, dityrosine and dopa (Menter et al., 2007).

A spectrofluorimetric method for the determination of boron on complexation of boric acid with ruthenium (II) has been developed. The complex on excitation at 360 nm, and at pH 8.9 emits at 600 nm (Nakano et al., 2008). The reaction of urea and boric acid has been used for the synthesis of BCNO nanocrystals containing carbon impurities. These crystals exhibit multicolor fluorescence under both single-photon and two-photon excitation (Liu et al., 2009a).

g. Emission Spectrometry

An inductively coupled plasma atomic emission spectrometric (ICP-AES) method has been developed for the determination of calcium gluconate (Ca), boric acid (B), phosphorous (P) and magnesium (Mg) in calcium borogluconate veterinary medicines. Reproducibility relative confidence intervals for a single sample are ± 1.4% (Ca), ± 1.8% (B), ± 2.6% (P) and ±1.4 % (Mg). The determined concentrations compare favorably with those of the alternative methods (Lyon and Spann, 1985). The application of ICP-AES to the determination of complex boron-containing compounds in biological tissue samples has been reported. The method gives a linear response for elemental boron concentration in the range of 0.05 to 100 ppm. It has also been applied to the determination of the boron content in compounds used for neutron capture therapy and in their biodistribution studies (Tamat et al., 1989). The suitability of boron-containing compounds in boron neutron capture therapy has been examined in two tumor models, a B16 pigmented melanoma and the RIFI sarcoma, in mice after different boron doses. Tissue and plasma levels of boron were measured using ICP-AES. The results showed that the proposed minimum effective tumor boron concentration of 15 ppm is achieved in both tumor models for the three compounds tested (Gregoire *et al.*, 1993). The emission of $GdAl_3(BO_3)_4$ polycrystals co-doped with Yb^{3+} and Eu^{3+} upon excitation at 465 nm occurs at 590, 613, 697 and 702 nm and is assigned to the electronic transition of $5D_0 \rightarrow 7FJ$ of Eu^{3+} ions (Yang *et al.*, 2009a).

A new method for analyzing the contents of several metallic elements in refractory tantalum-niobium slag by ICP-AES has been developed. The sample processing procedure involves the decomposition by potassium carbonate and boric acid at 950°C for 15 min. and leaching by hydrochloric and tartaric acids at 90°C for 30 min. the method showed RSD values between 0.27% and 5.48% and the recovery rates between 94.0% and 109.6% (Wang *et al.*, 2009a). The amount of leaching elements in the degradation of bioactive borosilicate glass scaffolds have been determined by ICP-AES. The study showed that borosilicate glass scaffold could be a promising candidate for bone tissue engineering material (Liu *et al.*, 2009b).

h. Laser-Induced Breakdown Spectroscopy

Laser-induced breakdown spectroscopy (LIBS) has been developed for determining the percentage of uranium in thorium-uranium mixed oxide pellets using boric acid as a binder. Calibration curves were established using U(II) at 263.553, 367.007, 447.233 and 454.363 nm emission lines. Except for 263.553 nm all the other emission lines exhibited a saturation effect due to self absorption when U amount exceeded 20 wt% in the Th-U mixture (Sarkar *et al.*, 2009).

i. X-Ray Photoelectron Spectroscopy

The surfaces of three imidazolium based ionic liquids with a common anion, 1-butyl-3-methylimidazolium, 1-hexyl-3-methylimidazolium and 1-octyl-3-methylimidazolium tetrafluoroborates, have been studied with angle-resolved X-ray photoelectron spectroscopy. Survey and high-resolution spectra were obtained at different take-off angles (0-84°) and no impurities were detected. The survey spectra at normal emission (0°) confirmed the stoichiometric composition of the liquids (Lockett *et al.*, 2008).

j. X-Ray Diffraction Analysis

Racemic amino alcohol derivatives can be resolved using boric acid and chiral 1,1'-bis-2-naphthol in solvents such as acetonitrile and methanol and the corresponding ammonium borate complexes identified by X-ray diffraction method (Periasamy *et al.*, 2001). The structure of the complex of urease, a Ni-containing metalloenzyme, with boric acid has been determined at 2.10 Å resolution. The complex shows the unprecedented binding mode of the competitive inhibitor to the dinuclear metal centre, with the boric acid molecule bridging the Ni ions and leaving in place the bridging hydroxide (Benini *et al.*, 2004). X-ray structures of the benzeneboronic acid and 2-phenylethaneboronic acid complexes of subtilisin BPN' (Novo) (Matthews *et al.*, 1975) and of the 2-phenylethaneboronic acid complex of α-chymotrypsin dimer (Tulinsky and Blevins, 1987) have been determined.

Single crystals of gadolinium triiron tetraborate (Klimin *et al.*, 2005), noncentrosymmetric borate (Reshak *et al.*, 2008), ammonium pentaborate (Balakrishnan *et al.*, 2008) and potassium pentaborate (Mary *et al.*, 2008), polyborate cluster compounds (Yang *et al.*, 2008a), heptaborate oxoanion isomer (Schubert *et al.*, 2008a), aluminum borate chloride (Gao *et al.*, 2008), complexes of copper (I) and silver (I) with bis(methimazolyl)borate and dihydrobis(2-mercaptothiazolyl) borate ligands (Beheshti *et al.*, 2008) wave-layered iron borate (Yang *et al.*, 2009b), strontium borate (Reshak *et al.*, 2009a), ternary borate oxide lead bismuth tetraoxide, (PbBiBO$_4$) (Reshak *et al.*, 2009b) and fluoroberyllium borates (McMillen *et al.*, 2009), have been characterized by X-ray diffraction.

A series of chiral ionic liquids which have either chiral cation (imidazolium group), chiral anion (borate ion) or both, have been synthesized. X-ray diffraction results confirm that the reaction to form the chiral spiral borate anion is stereospecific (Yu *et al.*, 2008a). Crystal structures of ionic liquid based mixtures of N,N-diethyl-N-2-methoxyethylammonium tetrafluoroborate have been determined by X-ray diffraction method. Most of the crystal structures in the mixtures are related to that of the pure compound, though unit cells in 1.1 and 6.1 M% benzene are different from the pure one (Imai *et al.*, 2009).

2.1.4.3. Electrochemical Techniques

a. Potentiometry

Hydroxyurea selective electrodes have been developed with beta-cyclodextrin used as ionophore and either tetrakis (*p*-chlorophenyl) borate (electrode 1) or tetrakis [3,4-bis (trifuoromethyl) phenyl] borate (electrode 2), as a fixed anionic site in a polymeric matrix of carboxylated polyvinyl chloride. Linear responses of hydroxyurea within a concentration range of 10^{-5} to 10^{-3} M at pH 3-6 were obtained by using these electrodes. The method has been applied to the determination of hydroxyurea in capsules and biological fluids (El-Kosasy, 2003). The electrochemical response characteristics of polymeric membrane sensors could be used for the potentiometric determination of zolpidem hemihydrate in tablets and biological fluids. The construction of the sensors is based on the formation of ion-pair complexes between the drug cation and ionic sites in the ratio of 1:2, respectively. A sensor has been fabricated by using 2,6-didodecyl-beta-cyclodextrin as the ionophore, polyurethane (Tecoflex) as a polymeric matrix and potassium tetraphenyl borate as the ionic site. It shows a linear response over the concentration range of 10^{-7} to 10^{-2} M (Kelani, 2004).

Liquid polymer membrane electrodes based on nickel and manganese phthalocyanines have been examined for use as anion-selective electrodes. The highest selectivity for perchlorate was observed for the electrode based on manganese phthalocyanine in the presence of the lipophilic anionic additive sodium tetrakis [3,5-bis(trifluoromethyl) phenyl] borate (Arvand et al., 2007). The cell toxicity of a widely used ionic liquid 1-butyl-3-methylimidazolium tetrafluoroborate to Chinese hamster lung fibroblast cells (V79 cell line) has been evaluated by potentiometric method. The study suggests that the ionic liquid could be used as a tool to control mammalian cell proliferation rate (Qiu and Zeng, 2008).

The effect of zinc and boric acid concentrations on the zinc transport properties though a cation exchange membrane (IONICS 67-HMR-412) has been evaluated by chronopotentiometry. The results show that the presence of boric acid produces the precipitation of zinc metaborate on the anodic layer of the cation-exchange membrane (Herraiz-Cardona et al., 2010).

b. Conductometry

Low frequency conductance titrations of bases with boric acid have been performed. Boric acid is a very weak acid and contributes to conductance only on the formation of borate ions in the titration (Eriksen, 1969). A novel method for the determination of the anions of boric acid, based on incompletely suppressed conductometric detection, has been developed. Within the linear range of detection, the correlation coefficient of the peak area of boric acid is 0.9985 (Huang et al., 1999).

c. Amperometry

An amperometric glucose biosensor with extended concentration range utilizing the complexation effect of borate has been developed. Borate buffer strongly decreases amperometric response of a glucose oxidase linked pO_2 or

H_2O_2 sensing electrode, extending substantially its linear calibration range. With increasing pH and concentration of the buffer the upper limit for glucose can be varied between 1 and 30 mM/L glucose (Macholan *et al.*, 1992). The effects of ionic liquids on enzymatic catalysis of glucose oxidase towards the oxidation of glucose depend on the nature of the ionic liquids which influence the electrocatalytic activity of the electrodes towards the oxidation of glucose (Wu *et al.*, 2009). Polyhedral boron clusters are proposed as new, chemically and biologically stable, redox labels for electrochemical DNA hybridization sensors. Selective detection of the redox labeled DNA-probe was achieved by means of covalently attached electroactive marker 7,8-dicarba-nido-undekaborate group. The above findings lead to the simultaneous detection of several DNA targets (Jelen *et al.*, 2009).

d. Voltammetry

The behavior of ciprofloxacin complex with copper at a mercury electrode has been investigated in borax-boric acid buffer by linear sweep voltammetry. In the presence of DNA, the formation of electrochemically non-active copper-DNA complexes results in the decrease of equilibrium concentration of copper and its peak current. The decrease of the peak current is proportional to DNA concentration (Zhang *et al.*, 2004).

Simple, rapid and reliable voltametric methods have been developed for the determination of zafirlukast in pharmaceutical formulations based on its electrochemical reduction at a hanging mercury drop electrode. Its electrochemical behavior was investigated in borate buffer (pH 8.0) using cyclic voltammetry and linear sweep voltammetry (Suslu and Altinoz, 2005). The direct electrochemistry of horseradish peroxidase immobilized in a chitosan-[C4mim][BF4] film has been studied by cyclic voltammetry on a glassy carbon electrode. The halfway potential of horseradish peroxidase has been found to be pH dependent, suggesting that a concomitant proton and electron transfer is occurring. The electrode kinetics and activation energies (14.20 kJ mol^{-1}) obtained are identical to those reported for horseradish peroxidase films in aqueous media (Long *et al.*, 2008).

2.1.4.4. Electrophoretic Techniques

a. Capillary Electrophoresis

A capillary zone electrophoresis (CZE) method has been developed for the simultaneous analysis of closely related polyhydroxy alkaloids (calystegines) through in situ complexation with borate ions. The method has been applied to the quantitative analysis of catystegines in plant extracts and the results have been confirmed by GC-MS (Daali *et al.*, 2000). Capillary electrophoresis (CE) has significantly higher efficiency, selectivity and speed than HPLC for the separation of a mixture of flavonoid-3-O-glycosides using boric acid as running buffer (pH 10.5). The migration order and the selectivity of the glycosides has been explained on the basis of in situ borate complexation of both the sugar moiety and the cis-1,2-hydroxyl groups on the flavonoid skeleton (Morin *et al.*, 1993). The electrophoretic mobility of small DNA fractions in Tris-borate-EDTA buffer compared to that in Tris-acetate-EDTA buffer appears to be due to the formation of highly charged deoxyribose-borate complexes (Strutz and Stellwagen, 1998). A capillary zone electrophoresis method has been used for the separation of glycerin-based polyols with a buffer solution containing 50% (v/v) acetonitrile and 10 mM sodium tetraborate. The method is applied to the determination of mono- and difunctional impurities in technical grade glycerin-based polyol products (Oudhoff *et al.*, 2004). Simultaneous determination of positional isomers of o-, m-, and p-benzenediols has been achieved by capillary zone electrophoresis with square wave amperometric detection using boric acid as running buffer. The method has been used to monitor the contents of benzenediol isomers in environmental waste water (Xie *et al.*, 2006).

Ion-association capillary zone electrophoresis has been applied to the separation and determination of unsaturated disaccharides of chondroitin sulfates and of oligosaccharides (tetra- and hexasaccharides) of hyaluronan using borate buffer modified by an ion-pairing reagent, tetrabutylammonium phosphate. The mechanism of separation is based on the interaction of anionic borate complexes with tetrabutylammonium ion inside the capillary tube. The method is used for the assay of chondroitin sulfates and hyaluronan in joint tissues whose relative abundance may vary in various diseases or after local treatment with antiinflammatory drugs (Payan *et al.*, 1998).

Several applications of capillary zone electrophoresis have been reported for the separation and determination of natural products and drugs including enantiomers of simendan (Li *et al.*, 2004), enantiomers of hydrobenzoin as borate complexes and structurally related compounds (Lin and Lin, 2004, Lin *et al.*, 2004a, 2004b, 2005a), aconite alkaloids (hypaconitine, aconitine and mesaconitine) in traditional Chinese and Tibetan medicines (Zhao *et al.*, 2004), saikosaponins a, c and d in Chinese herbal extracts of *Bupleurum chinese* DC (Lin *et al.*, 2005b), anthraquinones in Chinese herb *Paedicalyx attopevensis* Pierre ex Pitard (Qi *et al.*, 2004), active components of *Euphrasia regelii* (eukovoside, cinnamic acid and ferulic acid) (Shuya *et al.*, 2004), disaccharides and oligosaccharides of glycosaminoglycans coupled with an anionic fluorescent dye (Oonuki *et al.*, 2005), borate complexes of DL-pantothenic acid enantiomers (Aizawa *et al.*, 2006), bioactive flavone derivatives in Chinese herb *Seriphidium santolinum* Poljak (Qi *et al.*, 2006), isoflavones in red clover (*Trifolium pretense*) (Peng and Ye, 2006), flavonoids in *Lamiophlomis rotata* (Luo *et al.*, 2007), isoflavonoids in *Pueraria lobata* (Fang *et al.*, 2006), recombinant birch pollen allergen Bet v 1a preparations (Punzet *et al.*, 2006), kava lactones and flavonoid glycosides in *Scorzonera austriaca* (Jiang *et al.*, 2007), cetirizine dihydrochloride, paracetamol and phenylpropanolamine hydrochloride in tablets (Azhagvuel and Sekar, 2007), zafirlukast in pharmaceutical preparations (Suslu *et al.*, 2007), acidic non-steroidal anti-inflammatory drugs (Furlanetto *et al.*, 2007), derivatized mono- and disaccharides (Liu *et al.*, 2007), phenylpropanoids in Chinese medicine and its preparations (Li *et al.*, 2007), monosaccharides and derivatives (mannose, galacturonic acid, glucuronic acid, rhamnose, glucose, galactose, xylose, arabinose and fucose) (You *et al.*, 2008), mono-, di- and oligosaccharides (Chen *et al.*, 2009; Tseng *et al.*, 2009; Kazarian *et al.*, 2010), double-stranded DNA fragments (Yang *et al.*, 2007), segmented double-stranded RNA genome of rotavirus (Mishra *et al.*, 2010), supercoiled DNA (Ishido *et al.*, 2010), chiral vicinal diols (Liu *et al.*, 2008, 2009c), polyphenols (Jac *et al.*, 2008), anthracyclines (Whitaker *et al.*, 2008), basic and acidic proteins (lysozyme, cytochrome *c*, ribonuclease A, albumin and α-lactalbumin) (Wu *et al.*, 2008b; Kaneta *et al.*, 2009), edible nut seed proteins (Sathe *et al.*, 2009), carbohydrates (monosaccharides, oligosaccharides and glycosides) (Campa and Rossi, 2008), phenolic acids in *Majorana hortensis* (Petr *et al.*, 2008), aldoses (xylose, arabinose, glucose, rhamnose, fucose, galactose, mannose, glucuronic acid and galacturonic acid) (Yang *et al.*, 2008b), purines and pyrimidines (Haunschmidt *et al.*, 2008), biogenic amines (Zhang *et al.*, 2008b), phosphoamino acids (Deng *et al.*, 2008), phenol and m-nitrophenol (Wei *et al.*, 2008), phenols and nucleoside monophosphates (Tian *et al.*, 2008), 4-cyclic diarylheptanoids (rhoiptelol, juglanin A, juglanin B, juglanin C) and an alpha-tetralone derivative (sclerone) in green walnut husks (Li *et al.*, 2008), inorganic anions (chlorate, perchlorate, nitrate, nitrite and sulfate) and cations (ammonium and potassium) in explosive residues (Feng *et al.*, 2008), amino acids in *Sargassum fusiforme* (Chen *et al.*, 2005), dansylated amino acids enantiomers (Qi *et al.*, 2007, 2008; Qi and Yang, 2009a,b), a protein phosphatase (calcineurin) (Enayetul Babar *et al.*, 2008), terbutaline sulfate (Li *et al.*, 2009b), glucosamine in nutraceutical formulations after labeling with anthranilic acid (Volpi, 2009), fingerprints of *Citrus aurantium*

(Luo *et al.*, 2008), several drugs in pharmaceuticals and urine (Solangi *et al.*, 2009), enantiomers of basic drugs (Chen *et al.*, 2010), mercury and methylmercury (Deng *et al.*, 2009a), enumeration of *Lactobacillus delbrueckii* (subspecie bulgaricus) and *Streptococcus thermophilus* in yogurt (Lim *et al.*, 2009), cis-diol-containing compounds by borate complexation (Liu *et al.*, 2009d) and simultaneous determination of angiotensins II and 1-7 in plasma and urine of rats (Tenorio-Lopez *et al.*, 2010). These separations have been carried out using borate buffer for complexation or as an electrophoretic medium.

A simple and sensitive CE method has been reported for the separation and determination of zinc dimethyldithiocarbamate and zinc ethylenebisdithiocarbamate in boric acid buffer by direct UV detection at 254 nm. The detection limits of these compounds are 1.88×10^{-6} M and 2.48×10^{-6} M, respectively (Kumar Malik and Faubel, 2000). Ascorbic acid and sorbic acid have been separated by CZE in borate buffer and determined at 270 nm in fruit juices (Tang and Wu, 2005). A method for the determination of benzoyl peroxide in wheat flour has been developed by a micellar CE online sweeping concentration technique using borate as running buffer. The detection limit was 2 mg/L (Wang *et al.*, 2007).

Microchip capillary electrophoresis has been applied to the separation of monosaccharides of glycoproteins (Maeda *et al.*, 2006) and to follow the degradation of phenolic acid by aquatic plants (Ding and Garcia, 2006) using borate buffer. Chiral resolution of DL-pantothenic acid with a 1,3-diol structure (Kodama *et al.*, 2004a) and reducing monosaccharides (mannose, galactose, fucose, glucose, xylose and arabinose) (Kodama *et al.*, 2006, 2007) has been achieved using borate as a central ion of the chiral selector. An ionic liquid, 1-ethyl-3-methylimidazolium-tetrafluoroborate, has been used as the working electrolyte for chiral separation of complex enantiomers of dipeptides in glass microchip electrophoresis (Zeng *et al.*, 2007).

b. Capillary Isotachophoresis

Polyhydric alcohols, sorbitol and xylitol, are efficiently separated and determined by capillary isotachophoresis with conductometric detection. The method is based on the on-column complex formation equilibria between the polyols and boric acid using an operational system consisting of 10 mM HCl + 20 mM imidazole + 20 mM boric acid (pH 8.0). It has been applied to the simultaneous assay of borated sorbitol and xylitol in multicomponent infusion solutions (Pospisilova *et al.*, 1998). A method for isotachophoretic determination of sweeteners (acesulfame potassium, saccharine, aspartame, cyclamate, sorbitol, mannitol, lactitol and xylitol) in candies and chewing gums using conductometric detection has been reported. Boric acid is added to the aqueous sample solution to form borate complexes with polyhydroxyl compounds and make them migrate isotachophoretically. The calibration curves in the concentration range up to 2.5 mM are linear for all the components and the detection limits range between 0.024 and 0.081 mM (Herrmannova *et al.*, 2006).

c. Micellar Electrokinetic Chromatography (MEKC)

MEKC has been applied to the separation and determination of natural products and drugs such as enkephalin-related peptides derivatized using fluorescein isothiocyanin (Huang *et al.*, 2004), catechin and epicatechin enantiomers in tea drink (Kodama *et al.*, 2004b), anthraquinones (emodin, aloe-emodin and rhein) in Chinese teas (Zheng *et al.*, 2004), ephedrine and pseudoephedrine derivatized with 4-chloro-7-nitrobenzo-2-oxa-1, 3-diazol (Jianping *et al.*, 2005), mixture of penicillins (Puig *et al.*, 2005), corticosteroids (betamethsone, cortisone, prednisolone, 6-α-methylprednisolone, triamcinolone and prednisone) (Kuo and Wu, 2005), chiral aryl allenic acid enantiomers (Wang *et al.*, 2005), lignans in seeds of *Schisandra species* (Tian *et al.*, 2005), lamivudine and zidovudine in pharmaceutical formulations (Sekar and Azhaguvel, 2005), didanosine and impurities in bulk samples (Mallampati *et al.*, 2005), nitroaromatic and nitramine explosives in seawater samples (Giordano *et al.*, 2006), D-amino acids in fermented foods (Carlavilla *et al.*, 2006), chiral amino acids in conventional and transgenic maize (Herrero *et al.*, 2007), baicalein, baicalin and wogonin in *Scutellariae radix* with 1-butyl-3-methylimidazolium tetrafluoroborate ionic liquid as additive (Zhang *et al.*, 2006), traces of aminoglycoside antibiotics (kanamycin B, amikacin, neomycin B and paromomycin I) in bovine milk (Serrano and Silva, 2006), 1,4-benzodiazepines (alprazolam, bromazepam, chlordiazepoxide, diazepam, flunitrazepam, medazepam, oxazepam, nitrazepam) (Hancu *et al.*, 2007), arctiin and arctigenin in *Fructus arctii* and its herbal preparations (Lu *et al.*, 2007), bioactive constituents in Liuwei Dihuang pills (Zhao *et al.*, 2007), isoflavones in red clover (Zhang *et al.*, 2007b), resina draconis (Cao *et al.*, 2008), ciclopirox olamine in pharmaceutical formulations (Li *et al.*, 2008), oseltamivir in Tamiflu capsules (Jabbaribar *et al.*, 2008), gentamicin in antibiotic carriers (Kuhn *et al.*, 2008), nimesulide related compounds in pharmaceutical preparations (Zacharis *et al.*, 2009), ertapenem and its impurities in medicinal products (Michalska *et al.*, 2009), aliphatic amines

(Zhang *et al.*, 2009a), neutral glucosides (ginsenosides, ginsenoside and ginsenoside) (Cao *et al.*, 2009), budesonide and its impurities in pharmaceutical preparations (Furlanetto *et al.*, 2009), p-nitrobenzaldehyde and its product of Baylis-Hillman reaction, 2-[hydroxy(4-nitrophenyl)methyl]-2-cyclopenten-1-one (Qi *et al.*, 2009) and dioxouranium (VI), iron (III), copper (II) and nickel (II) using bis (salicylaldehyde) propylenediimine as chelating agent (Mirza *et al.*, 2008). In the MEKC separations borate buffers have been used as derivatization media or as a component of the running buffers (pH 8.0-10.0).

A new method has been developed for the sensitive and selective determination of short chain aliphatic amines in biological samples. The amines are converted into their N-hydroxysuccinimidyl fluorescein-O-acetate derivatives and measured by micellar electrokinetic capillary chromatography with laser-induced fluorescence detection. The amine derivatives were fully separated using 25 mM pH 9.6 boric acid electrolytes containing 60 mM sodium dodecyl sulphate. The detection limits were in the range of 0.02-0.1 nM (Deng *et al.*, 2009b).

2.1.5. Borate Interactions

The kinetics of borate interaction at ribose with NAD^+ and NMN^+ has been studied using the perturbation effect of borate on the addition of sulphite to the 4-position of nicotinamide ring. The rate of interaction at low borate buffer concentration is dependent on the concentration of both borate and boric acid (Johnson and Smith, 1976). Complexation of the ribose group of NAD^+ is preferred electrostatically over that of NADH resulting in the inhibition of this coenzyme system (Smith and Johnson, 1976). The direct measurement of boric acid and borate addition to NAD^+ and NADH by electrospray ionization mass spectrometry (ESI-MS) and ^{11}B NMR spectroscopy has been reported. The analysis shows that borate binds to both cis-2, 3-ribose diols on NAD^+ forming borate monoesters, borate diesters and diborate esters, whereas only borate monoesters are formed with NADH (Kim *et al.*, 2003). A mechanistic study of the complexation of boric acid with 4-isopropyl-tropolone has shown that the rate-determining step is the change in boron coordination from trigonal to tetrahedral form (Ishihara *et al.*, 1991).

The technique of multiple quanta filtered (MQF) NMR has been applied to the characterization of the binding of ligands containing quadrupolar nuclei to proteins. Relaxation time measurements indicate multiple binding of boric acid and borate ion to ferri- and ferrocytochrome c. The triple quantum relaxation of borate at a binding site is governed by dipolar interaction corresponding to an average B-H distance of 2.06 ± 0.07 A (Taler et al., 1999). The polysaccharide "schizophyllan (SPG)" has been shown to trap as-grown and cut single-walled carbon nanotubes (as-SWNTs and c-SWNTs, respectively). The c-SWNT-s-SPG (single stranded SPG) composites can be aligned using the covalent bond formation between boric acid derivatives and the 4,6-dihydroxyl group of the glucose side-chain unit (Tamesue et al., 2008). The self assembly of (bis(1-methyl-imidazol-2-yl)methyl)(1-methyl-4-nitroimidazol-2-yl)methyl)amine and boric acid results in a supramolecular structure containing bundled antiparallel imidazol-boric acid helices and boric acid filled one-dimensional channels (Cheruzel et al., 2005). Oxidized alginate and oxidized alginate blended with chitosan have been prepared in the presence of borax and calcium chloride, and their interactions with pyrimethanine, have been investigated using isothermal titration calorimetry. The enthalpy change of interaction has been found to be -4.86 ± 0.156 kJ mol^{-1} (Vieira et al., 2008). The electrostatic interaction between the ionic liquid 1-butyl-3-methylimidazolium tetrafluoroborate and calf thymus DNA has been studied by a surface electrochemical micromethod. The thermodynamic and kinetic parameters about the interactions such as the binding constant, the Gibbs energy of surface binding and the dissociation constant were obtained (Xie et al., 2008). The interactions of boric acid derivatives with oligo- and polyglucans has been studied by MS and NMR techniques. The formation of six-membered ring at C-4 and C-6 of the non-reducing glucose occurs in the case of monosaccahrides. An increase in the amout of the reagent results in the formation of a seven-membered ring at the secondary OH moieties (Meiland et al., 2010).

Borate anions, B $(OH)_4^-$ are known to associate with alkali and alkaline-earth metal cations in sea-water. The stability constants of the borate cation ion pairs of different metals (Na^+, Li^+, Mg^{2+}, Ca^{2+} and Sr^{2+}) have been determined by potentiometry (Rogers and van den Berg, 1988).

The interactions of poly-L-lysine coated boron nitride nanotubes (BNNTs) with C2C12 cells (muscle cells) have been investigated in terms of cytocompatability and BNNT internalization using confocal and transmission electron microscopy (Ciofani et al., 2010).

2.1.6. Flow Injection Analysis

A flow injection method for the determination of boron in ceramic materials involves spectrophotometric measurement of the decrease in pH produced by the reaction between boric acid and mannitol in the presence of an acid-base indicator. The relative standard deviations of the method are 0.7 and 0.4% for 4 and 8 μg/mL of B_2O_3, respectively. The limit of detection is 0.02 μg/mL of B_2O_3 (Sanchez-Ramos et al., 1998). The determination of calcium, magnesium and strontium in soils by flow injection flame atomic absorption spectrometry has been reported. Samples were dissolved in a mixture of hydrochloric and nitric acids and the digest obtained was evaporated to dryness to remove silicon. Boric acid was added to prevent the precipitation of the lanthanum releasing agent and determinations were made by the above method (Arslan and Tyson, 1999). A simple and rapid flow injection method has been reported for the determination of iron in blood serum after acid digestion. It is based on luminol chemiluminescence detection in the absence of added oxidant. The detection limit is 1.0 nmol/L with a sample throughput of 120 h^{-1}. The calibration graph is linear over the range 0.001-10 μmol/L with relative standard deviation of 3.2-5.0% (Waseem et al., 2004). Boron as boric acid in eye lotions has been determined by sequential injection analysis. The method is based on complexation between d-sorbitol and boric acid followed by the reaction with methyl orange and measurement of absorbance at 520 nm. The detection limit of boric acid is 0.06 mg / L (Van Staden and Tsanwani, 2002). The sensitivity improvement of a flow injection spectrophotometric method for the determination of ammonia has been examined based on an indophenol blue coloration reaction with salicylate and hypochloride in the presence of manganese (II) as a catalyst. Ammonia was absorbed into a boric acid solution and then analyzed by the proposed method (Tsuboi et al., 2002).

Flow injection analysis with electrospray ionization mass spectrometry has been used to investigate borate-nucleotide complex formation (Kim et al., 2004) (Section 1.2.5). A continuous online concentration method for the sensitive detection of alkaloids applying capillary electrophoresis-flow injection analysis with head-column field-amplified sample stacking has been developed using borate buffer. It has been applied to the determination of ephedrine and pseudoephedrine in pharmaceutical preparations with recoveries of 92.3-102.4% (Fan et al., 2005).

2.1.7. Thermal Analysis

Thermal analysis involves the measurement of a physical property of a substance as a function of controlled temperature increase. The technique has been applied to study the effect of sodium tetraborate on the thermal property of frozen aqueous sugar and polyol solutions. Addition of sodium tetraborate has been found to raise the thermal transition temperature of these solutions as a result of the formation of borate-sugar (mono-, di-, and oligosaccharides) and borate-polyol (dextran) complexes. The decreased solute mobility of borate complexes in frozen solutions suggests higher collapsible temperature in the freeze-drying process and improved stability of biological systems in frozen solutions (Izutsu et al., 2003). Thermal analysis of frozen solutions of mannitol containing sodium tetraborate has shown that the latter compound inhibits mannitol crystallization at sodium tetraborate / mannitol molar concentration ratios of approximately 0.05, which is much lower than that of boric acid. Inhibition of crystallization in frozen solutions results in highly amorphous mannitol in the freeze-dried solids. Changes in the thermal transition temperatures suggest reduced mannitol molecular mobility with an increase in the sodium tetraborate ratio (Izutsu et al., 2004).

Boric acid has been used to inhibit the crystallization of amorphous mannitol. At concentrations about 5% (w/w) of boric acid, the differential scanning calorimetry (DSC) scans indicate that a totally amorphous solid can be prepared by cooling the melted pre-mixture under ambient conditions. An increase in the glass transition temperature is observed with a corresponding increase in boric acid concentration. The rate of crystallization at 30°C for mannitol alone is 7000 times higher than that of mannitol containing 7.5% (w/w) of boric acid (Yoshinari et al., 2003).

The dielectric relaxation spectra of concentrated aqueous solutions of sucrose-borate mixtures have been measured in the supercooled and glassy regions in the frequency range of 40 Hz to 2 MHz. The secondary relaxation process was analyzed in the temperature range 183-283 K at water contents between 20 and 30 wt%. The relaxation times and the activation energy of the process were determined (Longinotti et al., 2008).

2.1.8. Water Analysis

Boron is widely distributed in surface and ground water predominantly as undissociated boric acid, and is found in ocean waters at a mean level of approximately 4.6 mg B/L. An analysis of US surface waters indicates that the 10th, 50th and 90th percentile B levels are 0.010, 0.070 and 0.387 mg B/L, respectively. The overall mean B concentration in Canadian surface waters in 1988 was reported as 0.16 to 0.20 mg B/L. High levels of B have been measured in human water supplies of northern Chile, with concentration ranging from 0.31 to 15.2 mg B/L. River water sources of B in UK and northern Italy range from 0.002 to 0.87 mg B/L, respectively, whereas German drinking waters have maximum B levels of 0.18 mg B/L. Bottled mineral water also represents a source of B exposure based on the largest survey of US and European products, with a mean level of 0.75 mg B/L and a range from < 0.005 to 4.35 mg B/L. Thus B intake from drinking water is highly variable and dependent on the geographic source of water (Coughlin, 1998). A preliminary review of boron data in the National Inorganic Radionuclide Survey by the Environmental Protection Agency (EPA) indicates the median boron level in US drinking water supplies to be 0.031 mg B/L (as boric acid and borax) and most exposures are less than 2.44 mg B/L (99th percentile). Thus the boron in US drinking water is safe for human consumption (Murray, 1995). The EPA health advisory committee recommends boron concentrations in drinking water not exceed 0.6 mg B/L (0.06 mM B) over a lifetime exposure (Moore, 1997). Drinking waters typically contain <1mg B/L, but the range is large and some population may be exposed to considerably more than 1 mg B/L (Richold, 1998).

The control of metal toxicity in waste waters including acid mine drainage has been achieved by the use of sulphate-reducing bacteria. Reuse of sulphate-reducing bacteria sludge and beads crosslinked with boric acid has been studied (Min et al., 2008).

The World Health Organization defines a boron level of 0.3 mg/L as the non-observed effect level (NOEL) for drinking water. High boron levels as boric acid in drinking water are toxic to humans (Polat et al., 2004). Safe concentrations of boron in irrigation water have been reported as 0.3 mg/L for sensitive plants, 1-2 mg/L for semitolerant plants and 2-4 mg/L for tolerant plants (Yilmaz et al., 2006). Several methods have been reported for the removal of boron from waters and waste waters. These include precipitation-coagulation, reverse osmosis, electrodialysis, solvent extraction, membrane filtration and adsorption (Ozturk and Kavak, 2005; Yazicigil and Oztekin, 2006). Mannitol doped and un-doped hybrid gels have been used for the removal of boron in drinking water. Mannitol is an effective complexing agent for boric acid and the hybrid gels were used to adsorb boric acid from water solutions (Liu et al., 2009e).

2.1.9. Clinical Analysis

A number of applications of analytical methods for the determination of boric acid and various drugs in biological materials have been reported. Boric acid in biological fluids has been determined spectrophotometrically by its reaction with protonated curcumin at 550 nm (Yoshida *et al.*, 1989b). The quantitative determination of percutaneous absorption of [10]B in [10]B-enriched boric acid and borates in biological materials has been made by inductively coupled plasma-mass spectrometry (Wester *et al.*, 1998a, 1998b, 1998c). Complex boronated compounds in tissues and plasmas have been determined by inductively coupled plasma atomic emission spectrometry (Tamat *et al.*, 1989; Gregoire *et al.*, 1993). The reaction of aminophylline with boric acid has been used for its spectrophotometric determination in mixed serum samples (Li *et al.*, 2009a). Rubomycin antibiotic has been determined by fluorescence measurements of its borate complex at 620 nm in blood, milk and urine samples (Alykov *et al.*, 1976).

Hydroxyurea can be determined in biological fluids by ion-selective potentiometry using different borate electrodes (El-Kosasy, 2003). A selective potentiometric determination of zolpidem hemitartrate in biological fluids using polymeric membrane electrodes with potassium tetraphenyl borate as the ionic site has been reported (Kelani, 2004). The complexation effect of borate has been utilized in amperometric glucose biosensors for measuring glucose levels in heparinized human blood or plasma samples (Macholan *et al.*, 1992).

A sensitive and selective method for the determination of fudosteine in human plasma has been developed. It involves a derivatization step with 9-fluorenylmethyl chloroformate in borate buffer and detection by HPLC-electrospray ionization mass spectrometry (ESI/MS). Excellent linearity has been obtained in the calibration range of 0.05-20 μg/mL fudosteine and the limit of detection is 0.03 μg/mL (Xu *et al.*, 2006).

A method for the determination of organic acids employing derivatization with trimethyloxonium tetrafluoroborate and sequential extraction by head space and direct immersion solid phase microextraction using gas chromatography-ion trap tandem mass spectrometry has been reported. It has been applied to the determination of furoic acid, hippuric acid, methylhippuric acid, mandelic acid, phenylglyoxylic acid and trans, trans mucanic acid directly in the urine. The automation of the method provides a number of advantages including reduced analysis time and greater reproducibility (Pacenti *et al.*, 2008).

Capillary zone electrophoresis with UV detection has been applied for the simultaneous determination of ascorbic acid and uric acid in human plasma (Zinellu *et al.*, 2004) and icariin and its metabolites (icaritiin and desmethylicaritin) in rat serum using sodium borate as a component of running buffer (Liu and Lou, 2004). The icariin data were used in pharmacokinetic studies.

Microchip electrophoresis has been used to determine glucose as glucose-borate complex in blood samples. The method is highly selective and sensitive with a RSD of 6.3-9.1% (Maeda *et al.*, 2007). Uric acid (Zhao *et al.*, 2008a) and α- and β-amanitin (Robinson-Fuentes *et al.*, 2008) have been determined in urine samples by capillary zone electrophoresis using borate buffer as background electrolyte.

A high throughput capillary electrophoresis method has been developed for the determination of plasma cysteinylglycine in retinal vein occlusion patients, using sodium phosphate/boric acid as an electrolytic solution (Zinellu *et al.*, 2008). Bromide (an anti-epileptic drug) in serum has been determined by capillary zone electrophoresis using a high viscosity running buffer (sodium tetraborate, sodium chloride and 25% glycerol, pH 9.24) and a short capillary (10 cm). The relative standard deviation values were ≤0.2% in terms of migration times and <2% in terms of peak areas (Pascali *et al.*, 2009).

Borate is widely used as a preservative in the clinical analysis of urine samples. It readily forms covalent adduct and reversible complexes with hydroxyl and carboxylate groups and the effects of borate preservation in ^{1}H NMR spectroscopy-based metabolic profiling of human urine samples have been assessed. The principal alterations in the ^{1}H resonance metabolite patterns have been observed for compounds such as mannitol, citrate and alpha-hydroxyisobutyrate and confirmed by ESI-MS analysis (Smith *et al.*, 2009). A sensitive capillary electrophoresis method for the determination of amino acids and catecholamines on reaction with 3-(4-bromobenzoyl)-2-quinolinecarboxaldehyde as a fluorogenic reagent has been developed. The optimal running buffer included boric acid and the limit of detection was found as low as 0.65 nM (Zhang *et al.*, 2009b).

2.1.10. Elemental Analysis

Boric acid has been used in the determination of certain elements as a chemical modifier. The determination of uranium in various geological materials like rocks, minerals, soils etc. is based on the reaction of samples with HF and HNO_3 at room temperature overnight. Boric acid is added to complex excess fluoride ions before the determination of uranium by laser fluorimetry (Ramdoss et al., 1997). In the determination of antimony by hydride generation technique a pretreatment procedure has been developed for the reduction of Sb (V) to Sb (III) in order to remove the effect of HF which strongly interferes with the reduction of antimony. It is based on the combined action of 1-cysteine and boric acid at 80°C. The pretreatment is effective in both HNO_3 and HCl media (D'Ulivo et al., 1998). A highly sensitive, selective and simple kinetic spectrophotometric method has been used for the determination of dissolved chromium species in natural and industrial waste water. It is based on the catalytic effect of Cr (III) and/or Cr (VI) on the oxidation of 2-amino-5-methylphenol with H_2O_2. Boric acid and Tween-40 exert pronounced activating and micellar sensitizing effects on the redox reaction, respectively (Mohamed et al., 2006). The use of boric acid in the determination of trace amount of indium in high purity antimony by electrothermal atomic absorption spectrometry has been reported. The HF used for the digestion of antimony interferes with the determination and can be eliminated by the addition of boric acid as a chemical modifier (Dash et al., 2006). For the determination of trace amounts of rubidium in high purity cesium chloride matrix by electrothermal atomic absorption spectrometry, the negative influence of the chloride matrix can not be eliminated. Due to the high dissociation energy of rubidium chloride, it is difficult to dissociate in the gas phase and is lost. Elimination of the interference is achieved by the addition of boric acid as a chemical modifier (Dash et al., 2007).

2.1.11. Borate Catalyzed / Inhibited Reactions

Boric acid catalyzes the selective esterification of alpha-hydroxycarboxylic acids. The procedure is high-yielding and applicable to alpha-hydroxycarboxylic acids in the presence of other carboxylic acids including beta-hydroxy acids within the same molecule (Houston et al., 2004). N-methyl-4-boronopyridinium iodide is a more effective catalyst than boric acid for the esterification of alpha-hydroxycarboxylic acids (Maki et al., 2005). The kinetics of the oxidation of substituted phenyl methyl sulfides by hydrogen peroxide in borate/boric acid buffers has been studied as a function of pH, total peroxide and total boron concentrations. Second-order rate constants at 25°C for the reaction of methyl-4-nitrophenyl sulfide and hydrogen peroxide, monoperoxoborate, or diperoxyborate are 8.29×10^{-5}, 1.51

$\times$ 10^{-2} and 1.06 $\times$ 10^{-2} M^{-1}s^{-1}, respectively (Davies *et al.*, 2005). An L-arabinose isomerase mutant enzyme has been used to catalyze the isomerization of D-galactose to D-tagatose with boric acid. Maximum production (74%) of D-tagatose occurs at pH 8.5-9.0, 60°C, and 0.4 molar ratio of boric acid to D-galactose (Lim *et al.*, 2007). The conversion yield of d-psicose from d-fructose by d-psicose epimerase increases with increasing molar ratios of borate to fructose up to a ratio of 0.6 (Kim *et al.*, 2008). The tributylborane-amine catalyst has been used to polymerize acrylate monomers (Sonnenschein *et al.*, 2009).

Selective *C*-methylation of phenol to *o*-cresol and 2,6-xylenol in high yields has been carried out with methanol over borate zirconia solid acid catalyst. The maximum conversion of 70 and 65% selectivity for corresponding *ortho*-alkylated products (*o*-cresol and 2,6-xylenol) is obtained with 1-2% anisole as *O*-alkylated product (Malshe *et al.*, 2004). Hydroxyborate anions act as novel anchors for zirconocene catalysts. These anions protonate the Zr-Me bond and afford a new route for the incorporation of a covalently bonded anionic functionality on organometallic complexes (Bibal *et al.*, 2008). An efficient methodology for the preparation of α-hydroxyamides via boric acid mediated addition of isonitriles onto aldehydes has been developed (Kumar *et al.*, 2010a). Various borate inhibitors have been tested for their inhibitory effects on soybean (Glycine max) urease. The K(i) values for boric acid, 4-bromophenylboronic acid, butylboronic acid and phenylboronic acid are 0.20 ± 0.05 mM, 0.22 ± 0.04 mM, 1.50 ± 0.10 mM and 2.00 ± 0.11 mM, respectively and the inhibition was competitive (Kumar and Kayastha, 2010).

The photodecomposition of hydrogen peroxide in borate/boric acid buffer is inhibited at higher pH. The quantum yield of the reaction is 0.8 ± 0.1 (Rey and Davies, 2006). Boric acid inhibits the acid-catalyzed depolymerization of cellulose in sulpholane at high temperature. Formation of dehydration products such as levoglucosenone, furfural and 5-hydroxymethyl furfural is also effectively inhibited (Kawamoto *et al.*, 2008). The inhibition of protease activity in rat liver chromatin by certain boronic acids has been reported (Carter *et al.*, 1977). The 2-phenylethaneboronic acid has been found to inhibit cell replication and this effect may be higher in rapidly proliferating cancer cells than in the normal tissues (Goz *et al.*, 1986).

2.1.12. Synthesis of Borate Compounds

An enantioselective synthesis of hermitamites A and B has been achieved utilizing a rhodium catalyzed conjugate addition of a chiral alkenyltrifluoroborate salt (Frost et al., 2008). Conjugate addition of potassium trifluoro (organo) borates to dehydroalanine derivatives mediated by a chiral rhodium catalyst and in situ enantioselective protonation leads to a variety of alpha-amino esters with high yields (Navarre et al., 2008). The preparation of stereocontrolled trisubstituted alkenes by using the rhodium catalyzed reaction of Baylis-Hillman adducts with either organoboronic acids or potassium trifluoroorganoborates has been reported (Gendrineau et al., 2009). The rhodium catalyzed addition of alkenyltrifluoroborates to both aromatic and aliphatic N-tert-butanesulfinyl aldimines in good yields (97%) and with very high diastereoselectivities (95:5 to >99:1) has been achieved (Brak and Ellman, 2009).

The petasis borono-mannich reaction and allylation of carbonyl compounds via transient allyl boronates generated by palladium-catalyzed substitution of allyl alcohols have been used as an efficient one-pot route to stereodefined alpha amino acids and homoallyl alcohols (Selander et al., 2007). Palladium catalyzed direct arylation of indoles (Zhao et al., 2008b), enaminones (Ge et al., 2008), C-H activation/aryl-aryl coupling of benzoic and phenyl acetic acids using aryltrifluoroborates (Wang et al., 2008) and reactions of alkynylborates with aryl halides (Ishida et al., 2009) have been reported. Palladium catalyzed 1,2-addition of aryl and alkenyltrifluoroborates to aldehydes using thioether-imidazolinium carbine ligands (Kuriyama et al., 2008) and cross-coupling reactions of alkenyltrifluoroborates with organic halides in aqueous media (Alacid and Najera, 2009) have been studied.

Alpha, beta-unstaurated aldehydes have been synthesized by the nickel-catalyzed 1,2-addition of arylborates (Sakurai et al., 2009). The photocatalytic activity of TiO_2 to the degradation of an azo-dye (Orange-G) is doubled in the presence of boric acid (Kwon et al., 2006). Boron doped CeO_2/TiO_2 mixed oxide photocatalyst have been prepared by adding boric acid and cerous nitrate during the hydrolyzation of titanium trichloride and tetrabutyl titanate. The photocatalytic activity increases with an increase in cerous nitrate concentration and decreases drastically due to its higher concentration (mol ratio of Ce/Ti >0.5) (Tang et al., 2006).

A large number of organic and inorganic borate compounds have been synthesized. These include Suzuki-Miyaura cross-coupling reactions of potassium heteroaryltrifluoroborates (Molander et al., 2009a), and potassium beta-trifluoroborato amides (Molander and Jean-Gerard, 2009), of cyclopropyl and cyclobutyltrifluoroborates with aryl and heteroaryl chlorides (Molander and Gormisky, 2008), of potassium trifluoroborates (Molander and Sandrock, 2009; Katz et al., 2009), of potassium trifluoroboratohomoenolates (Molander and Petrillo, 2008), of primary alkyltrifluoroborates with aryl chlorides (Dreher et al., 2009), of tetraarylborates with benzyl halides and 2-halopyridines (Bedford et al., 2009), of 1,1-dibromo- and 1,1-dichloro-1-alkenes (Chelucci, 2010), condensation reactions to form oxazoline-substituted potassium organotrifluoroborates (Molander et al., 2009b), aminomethylation of organotrifluoroborates (Molander et al., 2008a; Molander and Canturk, 2009), reductive amination of organotrifluoroborates (Molander and Cooper, 2008), regio- and stereoselective ring opening of enantiomerically enriched 2-aryl oxetanes and 2-aryl azetidines with aryl borates (Bertolini et al., 2008), cis-alkenyl pinacolboronates and potassium cis-alkenyltrifluoroborates (Molander and Ellis, 2008), crotylation of beta-branched alpha-methylaldehydes with potassium crotyltrifluoroborates (Tanaka et al., 2008) and synthesis of potassium alkoxymethyltrifluoroborates (Molander and Canturk, 2008), ethyltrifluoroborates (Molander et al., 2008b), aryl triolborates (Yu et al., 2008b), alcoholamine borates (Schubert et al., 2008b), aminodifluorosulfinium tetrafluoroborates (Beaulieu et al., 2009) fluorophenyl-substituted Fe-only hydrogenases (Wang et al., 2009b), borate-mixed crystals of the type of $(NH_4)KB_5O_8$ (Jesudurai et al., 2008), single crystals of $ReCa_4O(BO_3)_3$ for sensor applications (Zhang et al., 2008c), ionic liquids with mixed borate anions (Schreiner et al., 2009), (glycerol) borate based ionic liquids (Chiappe et al., 2010), aluminium nitride / boron nitride nanocomposites (Kusunose et al., 2008), $Mg_2B_2O_5$ nanowires (Tao and Li, 2008), copper borate with 14-ring channels (Yang et al., 2005), nickel polyborate (Ju et al., 2006), copper complexes of mono- and ditopic [(methylthio)methyl]borates (Ruth et al., 2008), sodium strontium pentaborate (Wu et al., 2008c), ferrocenylphosphonium borates (Ramos et al., 2009), diazoalkane adducts (Mankad and Peters, 2008), (pyrazol-1-yl) borates (Mutseneck et al., 2010), indium borates and rare earth substituted indium borates (Velchuri et al., 2009), potassium azidoaryltrifluoroborates (Cho et al., 2009), aluminoborates (Rong et al., 2009), iron borate (Yang et al., 2009b), fluoroberyllium borates (McMillen et al., 2009), Bi^{2+}-doped strontium borates (Peng and Wondraczek, 2009), mullite-type gallium borate (Cong et al., 2010), uranyl borates (Wang et al., 2010), fluoride borate (Zhang et al., 2010) and alkylzinc complexes with tris (triazolyl) borate (Kumar et al., 2010b).

REFERENCES

Acklov, S.Z., Burgers, P.C., McCarry, B.E., Terlouw, J.K. (1999). Structural analysis of diols by electrospray mass spectrometry on boric acid complexes. *Rapid Commun. Mass Spectom., 13*, 2406-2415.

Ahmad, I., Vaid, F.H.M. (2006). Photochemistry of flavins in aqueous and organic solvents, In: E. Silva, A.M. Edwards, Eds., *Flavins: Photochemistry and Photobiology*, Cambridge: Royal Society of Chemistry, pp. 13-40.

Ahmad, I., Ahmed, S., Sheraz, M.A., Vaid, F.H.M. (2008). Effect of borate buffer on the photolysis of riboflavin in aqueous solution. *J. Photochem. Photobiol. B: Biol., 93*, 82-87.

Ahmad, I., Ahmed, S., Sheraz, M.A., Vaid, F.H.M. (2009). Analytical applications of borates. *Mat. Sci. Res. J., 3,* 173-202.

Ahmad, I., Ahmed, S., Sheraz, M.A., Vaid, F.H.M. (2010). Borate: toxicity, effect on drug stability and analytical applications, In: M.P. Chung, Ed., *Handbook on Borates: Chemistry, Production and Applications*, New York, NY: Nova Science Publishers Inc., Chap. 2.

Aizawa, S., Yamamoto, A., Kodama, S. (2006). Mechanism of enantioseparation of DL-pantothenic acid in ligand exchange capillary electrophoresis using a diol-borate system. *Electrophoresis, 27*, 880-886.

Alacid, E., Nájera, C. (2009). Palladium-catalyzed cross-coupling reactions of potassium alkenyltrifluoroborates with organic halides in aqueous media. *J. Org. Chem., 74*, 2321-2327.

Alykov, N.M., Karibiants, M.A., Alykova, T.V. (1976). Fluorimetric determination of rubomycin with the boric acid reaction. *Antibiotiki, 21*, 920-921.

Anderson, T.N., Lucht, R.P., Priyadarsan, S., Annamalai, K., Caton, J.A. (2007). In situ measurements of nitric oxide in coal-combustion exhaust using a sensor based on a widely tunable external-cavity GaN diode laser. *Appl. Opt., 46*, 3946-3957.

Ankli, A., Reich, E., Steiner, M. (2008). Rapid high-performance thin-layer chromatographic method for detection of 5% adulteration of black cohosh with *Cimicifuga foetida, C. heracleifolia, C. dahurica*, or *C. americana. J. AOAC Int., 91*, 1257-1264.

Arslan, Z., Tyson, J.F. (1999). Determination of calcium, magnesium and strontium in soils by flow injection flame atomic absorption spectrometry. *Talanta, 50*, 929-937.

Arvand, M., Pourhabib, A., Shemshadi, R., Giahi, M. (2007). The potentiometric behavior of polymer-supported metallophthalocyanines used as anion-selective electrodes. *Anal. Bioanal. Chem.*, *387*, 1033-1039.

Azhagvuel, S., Sekar, R. (2007). Method development and validation for the simultaneous determination of cetirizine dihydrochloride, paracetamol, and phenylpropanolamine hydrochloride in tablets by capillary zone electrophoresis. *J. Pharm. Biomed. Anal.*, *43*, 873-878.

Aznarez, J., Ferrer, A., Rabadan, J.M., Marco, L. (1985). Extractive spectrophotometric and fluorimetric determination of boron with 2,2,4-trimethyl-1,3-pentanediol and carminic acid. *Talanta*, *32*, 1156-1158.

Balakrishnan, T., Bhagavannarayana, G., Ramamurthi, K. (2008). Growth, structural, optical, thermal and mechanical properties of ammonium pentaborate single crystal. *Spectrochim. Acta A Mol. Biomol. Spectrosc.*, *71*, 578-583.

Balasubramanian, R., Lakshmi Narasimhan, T.S., Viswanathan, R., Nalini, S. (2008). Investigation of the vaporization of boric acid by transpiration thermogravimetry and knudsen effusion mass spectrometry. *J. Phys. Chem. B.*, *112*, 13873-13884.

Beaulieu, F., Beauregard, L.P., Courchesne, G., Couturier, M., LaFlamme, F., L'Heureux, A. (2009). Aminodifluorosulfinium tetrafluoroborate salts as stable and crystalline deoxofluorinating reagents. *Org. Lett.*, *11*, 5050-5053.

Bedford, R.B., Hall, M.A., Hodges, G.R., Huwe, M., Wilkinson, M.C. (2009). Simple mixed Fe-Zn catalysts for the Suzuki couplings of tetraarylborates with benzyl halides and 2-halopyridines. *Chem. Commun. (Camb)*, *42*, 6430-6432.

Beheshti, A., Clegg, W., Nobakht, V., Panahi Mehr, M., Russo, L. (2008). Complexes of copper(I) and silver(I) with bis(methimazolyl)borate and dihydrobis(2-mercaptothiazolyl)borate ligands. *Dalton Trans.*, *14*, 6641-6646.

Benini, S., Rypniewski, W.R., Wilson, K.S., Mangani, S., Ciurli, S. (2004). Molecular details of urease inhibition by boric acid: insights into the catalytic mechanism. *J. Am. Chem. Soc.*, *126*, 3714-3715.

Bertolini, F., Crotti, S., Di Bussolo, V., Macchia, F., Pineschi, M. (2008). Regio- and stereoselective ring opening of enantiomerically enriched 2-aryl oxetanes and 2-aryl azetidines with aryl borates. *J. Org. Chem.*, *73*, 8998-9007.

Bibal, C., Santini, C.C., Chauvin, Y., Vallée, C., Olivier-Bourbigou, H. (2008). A selective synthesis of hydroxyborate anions as novel anchors for zirconocene catalysts. *Dalton Trans.*, *7*, 2866-2870.

Bishop, M., Shahid, N., Yang, J., Barron, A.R. (2004). Determination of the mode and efficacy of the cross-linking of guar by borate using MAS ^{11}B NMR of borate cross-linked guar in combination with solution ^{11}B NMR of model systems. *Dalton Trans.*, *17*, 2621-2634.

Block J.H., Roche, E.B., Soine, T.O., Wilson, C.O. (1974). *Inorganic Medicinal and Pharmaceutical Chemistry*, Philadelphia, PA: Lea & Febiger, pp. 142-146.

Bornhorst, J.A., Hunt, J.W., Urry, F.M., McMillin, G.A. (2005). Comparison of sample preservation methods for clinical trace element analysis by inductively coupled plasma mass spectrometry. *Am. J. Clin. Pathol.*, *123*, 578-583.

Brak, K., Ellman, J.A. (2009). Asymmetric synthesis of alpha-branched allylic amines by the Rh(I)-catalyzed addition of alkenyltrifluoroborates to N-tert-butanesulfinyl aldimines. *J. Am. Chem. Soc.*, *131*, 3850-3851.

Bravo, M., Lespes, G., De Gregori, I., Pinochet, H., Potin-Gautier, M. (2004). Identification of sulfur interferences during organotin determination in harbour sediment samples by sodium tetraethyl borate ethylation and gas chromatography-pulsed flame photometric detection. *J. Chromatogr. A.*, *1046*, 217-224.

British Pharmacopoeia (2009). London: Her Majesty's Stationary Office, Electronic Version.

Campa, C., Rossi, M. (2008). Capillary electrophoresis of neutral carbohydrates: mono- oligosaccharides, glycosides. *Methods Mol. Biol.*, *384*, 247-305.

Cao, J., Li, B., Chang, Y.X., Li, P. (2009). Direct on-line analysis of neutral analytes by dual sweeping via complexation and organic solvent field enhancement in nonionic MEKC. *Electrophoresis*, *30*, 1372-1379.

Cao, Y., Gong, W., Li, N., Yin, C., Wang, Y. (2008). Comparison of microemulsion electrokinetic chromatography with high-performance liquid chromatography for fingerprint analysis of resina draconis. *Anal. Bioanal. Chem.*, *392*, 1003-1010.

Carlavilla, D., Moreno-Arribas, M.V., Fanali, S., Cifuentes, A. (2006). Chiral MEKC-LIF of amino acids in foods: analysis of vinegars. *Electrophoresis*, *27*, 2551-2557.

Carlson M., Thompson, R.D. (1998). Determination of borates in caviare by ion-exclusion chromatography. *Food Addit. Contam.*, *15*, 898-905.

Carretti, E., Grassi, S., Cossalter, M., Natali, I., Caminati, G., Weiss, R.G., Baglioni, P., Dei, L. (2009). Poly(vinyl alcohol)-borate hydro/cosolvent gels: viscoelastic properties, solubilizing power, and application to art conservation. *Langmuir, 25*, 8656-8662.

Carter, D.B., Ross, D.A., Ishaq, K.S., Suarez, G.M., Chae, C.B. (1977). The inhibition of rat liver chromatin protease by congeners of the phenylboronic acids. *Biochim. Biophys. Acta, 484*, 103-108.

Chelucci, G. (2010). Suzuki-Miyaura reactions of homo *gem*-dihalovinyl systems, In: M. P. Chung, Ed., *Handbook on Borates, Chemistry, Production and Applications*, NewYork, NY: Nova Science Publishers, Inc., Chap. 7.

Chen, B., Du, Y., Wang, H. (2010). Study on enantiomeric separation of basic drugs by NACE in methanol-based medium using erythromycin lactobionate as a chiral selector. *Electrophoresis, 31*, 371-377.

Chen, F., Wang, S., Guo, W., Hu, M. (2005). Determination of amino acids in Sargassum fusiforme by high performance capillary electrophoresis. *Talanta, 66*, 755-761.

Chen, J., He, L., Abo, M., Zhang, J., Sato, K., Okubo, A. (2009). Influence of borate complexation on the electrophoretic behavior of 2-AA derivatized saccharides in capillary electrophoresis. *Carbohydr. Res., 344*, 1141-1145.

Cheruzel, L.E., Mashuta, M.S., Buchanan, R.M. (2005). A supramolecular assembly of side-by-side polyimidazole tripod coils stabilized by pi-pi stacking and unique boric acid templated hydrogen bonding interactions. *Chem. Commun., 7*, 2223-2225.

Chiappe, C., Signori, F., Valentini, G., Marchetti, L., Pomelli, C.S., Bellina, F. (2010). Novel (Glycerol)borate-based ionic liquids: An experimental and theoretical study. *J. Phys. Chem. B*, doi: 10.1021/jp100809x, in press.

Cho, Y.A., Kim, D.S., Ahn, H.R., Canturk, B., Molander, G.A., Ham, J. (2009). Preparation of potassium azidoaryltrifluoroborates and their cross-coupling with aryl halides. *Org. Lett., 11*, 4330-4333.

Ciofani, G., Ricotti, L., Danti, S., Moscato, S., Nesti, C., D'Alessandro, D., Dinucci, D., Chiellini, F., Pietrabissa, A., Petrini, M., Menciassi, A. (2010). Investigation of interactions between poly-l-lysine-coated boron nitride nanotubes and C2C12 cells: up-take, cytocompatibility, and differentiation. *Int. J. Nanomed., 5*, 285-298

Cong, R., Yang, T., Li, K., Li, H., You, L., Liao, F., Wang, Y., Lin, J. (2010). Mullite-type $Ga_4B_2O_9$: structure and order-disorder phenomenon. *Acta Crystallogr. B, 66*, 141-150.

Coughlin, J.R. (1998). Sources of human exposure: overview of water supplies as sources of boron. *Biol. Trace Elem. Res., 66*, 87-100.

D'Ulivo, A., Lampugnani, L., Faraci, D., Tsalev, D.L., Zamboni, A. (1998). Elimination of hydrofluoric acid interference in the determination of antimony by the hydride generation technique. *Talanta, 45*, 801-806.

Daali, Y., Bekkouche, K., Cherkaoui, S., Christen, P., Veuthey, J.L. (2000). Use of borate complexation for the separation of non-UV-absorbing calystegines by capillary electrophoresis. *J. Chromatogr. A, 903*, 237-244.

Dash, K., Thargavel, S., Chaurasia, S.C., Arunachalam, J. (2006). Determination of indium in high purity antimony by electrothermal atomic absorption spectrometry (ETAAS) using boric acid as a modifier. *Talanta, 70*, 602-608.

Dash, K., Thargavel, S., Chaurasia, S.C., Arunachalam, J. (2007). Determination of traces of rubidium in high purity cesium chloride by electrothermal atomic absorption spectrometry (ETAAS) using boric acid as a modifier. *Anal. Chim. Acta, 584*, 210-214.

Davidson, A.G. (1984). Difference spectrophotometric assay of 1,2-diphenolic drugs in pharmaceutical formulations-I. Boric acid reagent. *J. Pharm. Biomed. Anal., 2*, 45-52.

Davies, D.M., Deary, M.E., Ouill, K., Smith, R.A. (2005). Borate-catalyzed reactions of hydrogen peroxide: kinetics and mechanism of the oxidation of organic sulphides by peroxoborates. *Chemistry, 11*, 3552-3558.

De Muynck, C., Beauprez, J., Soetaert, W., Vandamme, E.J. (2006). Boric acid as a mobile phase additive for high performance liquid chromatography separation of ribose, arabinose and ribulose. *J. Chromatogr. A, 1101*, 115-121.

Deng, Y.H., Zhang, H.S., Wang, H. (2008). Rapid and sensitive determination of phosphoamino acids in phosvitin by N-hydroxysuccinimidyl fluorescein-O-acetate derivatization and capillary zone electrophoresis with laser-induced fluorescence detection. *Anal. Bioanal. Chem., 392*, 231-238.

Deng, Y.H., Wang, H., Zhong, L., Zhang, H.S. (2009b). Trace determination of short-chain aliphatic amines in biological samples by micellar electrokinetic capillary chromatography with laser-induced fluorescence detection. *Talanta, 77*, 1337-1342.

Deng, B., Xiao, Y., Xu, X., Zhu, P., Liang, S., Mo, W. (2009a). Cold vapor generation interface for mercury speciation coupling capillary electrophoresis with electrothermal quartz tube furnace atomic absorption spectrometry: determination of mercury and methylmercury. *Talanta*, *79*, 1265-1269.

Dhavile, S.M., Shekhar, R., Thangavel, S., Chaurasia, S.C., Arunachalam, J. (2008). Determination of trace phosphorus in zirconium-niobium alloy and Zircaloy by UV-vis spectrophotometry. *Talanta*, *76*, 134-137.

Ding, Y., Garcia, C.D. (2006). Application of microchip-CE electrophoresis to follow the degradation of phenolic acids by aquatic plants. *Electrophoresis*, *27*, 5119-5127.

Dreher, S.D., Lim, S.E., Sandrock, D.L., Molander, G.A. (2009). Suzuki-Miyaura cross-coupling reactions of primary alkyltrifluoroborates with aryl chlorides. *J. Org. Chem.*, *74*, 3626-3631.

Edwards, J.O., Morrison, G.C., Ross, V.F., Schultz, J.W. (1955). The structure of the aqueous borate ion. *J. Am. Chem. Soc.*, *77*, 266-268.

El-Kosasy, A.M. (2003). Determination of hydroxyurea in capsules and biological fluids by ion-selective potentiometry and fluorimetry. *J. AOAC Int.*, *86*, 15-21.

Enayetul Babar, S.M., Song, E.J., Yoo, Y.S. (2008). Analysis of calcineurin activity by capillary electrophoresis with laser-induced fluorescence detection using peptide substrate. *J. Sep. Sci.*, *31*, 579-587.

Eriksen, S. (1969). Current flow methods, In: L.G. Chatten, Ed., *Pharmaceutical Chemistry*, New York, NY: Marcel Dekker, pp. 527-530.

Fan, L., Cheng, Y., Li, Y., Chen, H., Chen, X. Hu, Z. (2005). Head-column field-amplified sample stacking in a capillary electrophoresis-flow injection system. *Electrophoresis*, *26*, 4345-4354.

Fang, C., Wan, X., Tan, H., Jiang, C. (2006). Separation and determination of isoflavonoids in several kudzu samples by high-performance capillary electrophoresis (HPCE). *Ann. Chim.*, *96*, 117-124.

Feng, J., Guo, B., Lin, J.M., Xu, J., Zhou, H., Sun, Y., Liu, Y., Quan, Y., Lu, X. (2008). Determination of inorganic ions in explosive residues by capillary zone electrophoresis. *Se Pu*, *26*, 667-671.

Fine, J.B., Sprecher, H. (1982). Unidimensional thin-layer chromatography of phospholipids on boric acid-impregnated plates. *J. Lipid Res.*, *23*, 660-663.

Freire, F., Quiñoá, E., Riguera, R. (2008). In tube determination of the absolute configuration of alpha- and beta-hydroxy acids by NMR via chiral BINOL borates. *Chem. Commun. (Camb.)*, *21*, 4147-4149.

Frost, C.G., Penrose, S.D., Gleave, R. (2008). Rhodium catalysed conjugate addition of a chiral alkenyltrifluoroborate salt: the enantioselective synthesis of hermitamides A and B. *Org. Biomol. Chem.*, *6*, 4340-4347.

Fuguet, E., Reta, M., Gibert, C., Roses, M., Bosch, E., Rafols, C. (2008). Critical evaluation of buffering solutions for pKa determination by capillary electrophoresis. *Electrophoresis*, *29*, 2841-2851.

Furlanetto, S., Orlandini, S., Giannini, I., Beretta, G., Pinzauti, S. (2009). Pitfalls and success of experimental design in the development of a mixed MEKC method for the analysis of budesonide and its impurities. *Electrophoresis*, *30*, 633-643.

Furlanetto, S., Lanteri, S., Orlandini, S., Gotti, R., Giannini, I., Pinzauti, S. (2007). Selection of background electrolyte for CZE analysis by a chemometric approach. Part I. Separation of a mixture of acidic non-steroidal anti-inflammatory drugs. *J. Pharm. Biomed. Anal.*, *43*, 1388-1401.

Gao, W., Wang, Y., Li, G., Liao, F., You, L., Lin, J. (2008). Synthesis and structure of an aluminum borate chloride consisting of 12-membered borate rings and aluminate clusters. *Inorg. Chem.*, *47*, 7080-7082.

Gaspar, A., Schmitt-Kopplin, P. (2010). Letter: Identification of polyborate ions in aqueous solution by electrospray ionisation Fourier transform ion cyclotron resonance mass spectrometry. *Eur. J. Mass Spectrom.*, *16*, 237-242.

Gaspar, A., Harrir, M., Lucio, M., Hertkorn, N., Schmitt-Kopplin, P. (2008). Targeted borate complex formation as followed with electrospray ionization Fourier transform ion cyclotron mass spectrometry: monomolecular model system and polyborate formation. *Rapid Commun. Mass Spectrom.*, *22*, 3119-3129.

Ge, H., Niphakis, M.J., Georg, G.I. (2008). Palladium(II)-catalyzed direct arylation of enaminones using organotrifluoroborates. *J. Am. Chem. Soc.*, *130*, 3708-3709.

Gendrineau, T., Demoulin, N., Navarre, L., Genet, J.P., Darses, S. (2009). Rhodium-catalyzed formation of stereocontrolled trisubstituted alkenes from Baylis-Hillman adducts. *Chemistry*, *15*, 4710-4715.

Giordano, B.C., Copper, C.L., Collins, G.E. (2006). Micellar electrokinetic chromatography and capillary electrochromatography of nitroaromatic explosives in seawater. *Electrophoresis*, *27*, 778-786.

Goz, B., Ganguli, C., Troeonis, M., Wyrick, S., Ishaq, K.S., Katzenellenbogen, J.A. (1986). Compounds that inhibit chymotrypsin and cell replication. *Biochem. Pharmacol.*, *35*, 3587-3591.

Gregoire, V., Begg, A.C., Huiskamp, R., Veiryk, R., Bartelink, H. (1993). Selectivity of boron carriers for boron neutron capture therapy:

pharmacological studies with borocapture sodium, L-boronophenylalanine and boric acid in murine tumors. *Radiother. Oncol.*, *27*, 46-54.

Guerrero, L., Martínez-Olondris, P., Rigol, M., Esperatti, M., Esquinas, C., Luque, N., Piner, R., Torres, A., Soy, D. (2010). Development and validation of a high performance liquid chromatography method to determine linezolid concentrations in pig pulmonary tissue. *Clin. Chem. Lab. Med.*, *48*, 391-398.

Habicht, S.C., Vinueza, N.R., Archibold, E.F., Duan, P., Kenttämaa, H.I. (2008). Identification of the carboxylic acid functionality by using electrospray ionization and ion-molecule reactions in a modified linear quadrupole ion trap mass spectrometer. *Anal Chem.*, *80*, 3416-3421.

Hamano-Nagaoka, M., Matsuda, R., Maitani, T. (2008). Development and evaluation of determination methods for boric acid in agar using ICP-AES and ICP-MS. *Shokuhin Eiseigaku Zasshi*, *49*, 333-338.

Hancu, G., Gáspár, A., Gyéresi, A. (2007). Separation of 1,4-benzodiazepines by micellar elektrokinetic capillary chromatography. *J. Biochem. Biophys. Methods.*, *69*, 251-259.

Haunschmidt, M., Buchberger, W., Klampfl, C.W. (2008). Investigations on the migration behaviour of purines and pyrimidines in capillary electromigration techniques with UV detection and mass spectrometric detection. *J. Chromatogr. A.*, *1213*, 88-92.

Hepp, N.M., Mindak, W.R., Cheng, J. (2009). Determination of total lead in lipstick: development and validation of a microwave assisted digestion, inductively coupled plasma mass spectrometric method. *J. Cosmet. Sci.*, *60*, 405-414.

Herraiz-Cardona, I., Ortega, E., Perez-Herranz, V. (2010). Evaluation of the Zn^{2+} transport properties through a cation-exchange membrane by chronopotentiometry. *J. Colloid Interface Sci.*, *341*, 380-385.

Herrero, M., Ibáñez, E., Martín-Alvarez, P.J., Cifuentes, A. (2007). Analysis of chiral amino acids in conventional and transgenic maize. *Anal. Chem.*, *79*, 5071-5077.

Herrmannova, M., Krivankova, L., Bartos, M., Vytras, K. (2006). Direct simultaneous determination of eight sweeteners in food by capillary isotachophoresis. *J. Sep. Sci.*, *29*, 1132-1137.

Higa, S., Kishimoto, S. (1986). Isolation of 2-hydroxycarboxylic acids with a boronate affinity gel. *Anal. Biochem.*, *154*, 71-74.

Higashi, Y., Watanabe, N., Sasaki, T., Fujii, Y. (2003a). Measurement of beta-methyl digoxin level in serum from patients by enzyme immunoassay using novel specific antiserum with a phenylboric acid column. *Ther. Drug Monit., 25,* 452-456.

Higashi, Y., Watanabe, N., Sasaki, T., Fujii, Y. (2003b). Pharmacokinetic study of beta-methyldigoxin by enzyme immunoassay using a novel specific antiserum in rats. *Biol. Pharm. Bull., 26,* 247-251.

Houston, T.A., Wilkinson, B.L., Blanchfield, J.T. (2004). Boric acid catalyzed chemoselective esterification of alpha-hydoxycarboxylic acids. *Org. Lett., 6,* 679-681.

Hu, S., Chen, Y., Zhu, H., Zhu, J., Yan, N., Chen, X. (2009). In situ synthesis of di-n-butyl l-tartrate-boric acid complex chiral selector and its application in chiral microemulsion electrokinetic chromatography. *J. Chromatogr. A., 1216,* 7932-7940.

Huang, Y., Mou, S., Hou, X. Liu, K. (1999). The determination of anions of very weak acids by incompletely suppressed conductometric detection. *Se Pu, 17,* 287-289.

Huang, Y., Duan, J., Chen, Q., Chen, G. (2004). Micellar electrokinetic chromatography of enkephalin-related peptides with laser-induced fluorescence detection. *Electrophoresis, 25,* 1051-1057.

Imai, Y., Abe, H., Yoshimura, Y. (2009). X-ray diffraction study of ionic liquid based mixtures. *J. Phys. Chem. B., 113,* 2013-2018.

Imai, Y., Ito, S., Maruta, K., Fujita, K. (1988). Simultaneous determination of catecholamines and serotonin by liquid chromatography, after treatment with boric acid gel. *Clin. Chem., 34,* 528-530.

Ishida, N., Shimamoto, Y., Murakami, M. (2009). Stereoselective synthesis of (E)-(trisubstituted alkenyl) borinic esters: stereochemistry reversed by ligand in the palladium-catalyzed reaction of alkynylborates with aryl halides. *Org. Lett., 11,* 5434-5437.

Ishido, T., Ishikawa, M., Hirano, K. (2010). Analysis of supercoiled DNA by agarose gel electrophoresis using low-conducting sodium threonine medium. *Anal. Biochem., 400,* 148-150.

Ishihara, K., Mouri, Y., Funahashi, S., Tanaka, M. (1991). Mechanistic study of the complex formation of boric acid. *Inorg. Chem., 30,* 2356-2360.

Ito, R., Kawaguchi, M., Sakui, N., Okanouchi, N., Saito, K., Seto, Y., Nakazawa, H. (2009). Stir bar sorptive extraction with in situ derivatization and thermal desorption-gas chromatography-mass spectrometry for trace analysis of methylmercury and mercury(II) in water sample. *Talanta, 77,* 1295-1298.

Izutsu, K., Rimando, A., Aoyagi, N., Kojima, S. (2003). Effect of sodium tetraborate (borax) on the thermal properties of frozen aqueous sugar and polyol solutions. *Chem. Pharm. Bull. (Tokyo)*, *51*, 663-666.

Izutsu, K., Ocheda, S.O., Aoyagi, N., Kojima, S. (2004). Effect of sodium tetraborate and boric acid on nonisothermal mannitol crystallization in frozen solutions and freeze-dried solids. *Int. J. Pharm.*, *273*, 85-93.

Jabbaribar, F., Mortazavi, A., Jalali-Milani, R., Jouyban, A. (2008). Analysis of oseltamivir in Tamiflu capsules using micellar electrokinetic chromatography. *Chem. Pharm. Bull.*, *56*, 1639-1644.

Jác, P., Polásek, M., Batista, A.I., Kaderová, L. (2008). Tungstate as complex-forming reagent facilitating separation of selected polyphenols by capillary electrophoresis and its comparison with borate. *Electrophoresis*, *29*, 843-851.

Jelen, F., Olejniczak, A.B., Kourilova, A., Lesnikowski, Z.J., Palecek, E. (2009). Electrochemical DNA detection based on the polyhedral boron cluster label. *Anal. Chem.*, *81*, 840-844.

Jeon, Y., Sung, J., Kim, D., Seo, C., Cheong, H., Ouchi, Y., Ozawa, R., Hamaguchi H.O. (2008b). Structural change of 1-butyl-3-methylimidazolium tetrafluoroborate + water mixtures studied by infrared vibrational spectroscopy. *J. Phys. Chem. B.*, *112*, 923-928.

Jeon, Y., Sung, J., Seo, C., Lim, H., Cheong, H., Kang, M., Moon, B., Ouchi, Y., Kim D. (2008a). Structures of ionic liquids with different anions studied by infrared vibration spectroscopy. *J. Phys. Chem. B.*, *112*, 4735-4740.

Jesudurai, J.G., Prabha, K., Christy, P.D., Madhavan, J., Sagayaraj, P. (2008). Synthesis, growth and characterization of new borate-mixed crystals of type $(NH_4)_{1-x}K_xB_5O_8$. *Spectrochim Acta A Mol. Biomol. Spectrosc.*, *71*, 1371-1378.

Jeyakumar, S., Raut, V.V., Ramakumar, K.L. (2008). Simultaneous determination of trace amounts of borate, chloride and fluoride in nuclear fuels employing ion chromatography (IC) after their extraction by pyrohydrolysis. *Talanta*, *76*, 1246-1251.

Jiang, T.F., Wang, Y.H., Lv, Z.H., Yue, M.E. (2007). Determination of kava lactones and flavonoid glycoside in *Scorzonera austriaca* by capillary zone electrophoresis. *J. Pharm. Biomed. Anal.*, *43*, 854-858.

Jianping, X., Jiyou, Z., Jiaqin, L., Jianniao, T., Xingguo, C., Zhide, H. (2005). Rapid and sensitive determination of ephedrine and pseudoephedrine by micellar electrokinetic chromatography with on-line regenerating covalent coating. *Biomed. Chromatogr.*, *19*, 9-14.

Johnson, S.L., Smith, K.W. (1976). The interaction of borate and sulphite with pyridine nucleotides. *Biochemistry, 15*, 553-559.

Ju, J., Sasaki, J., Yang, T., Kasamatsu, S., Negishi, E., Li, G., Lin, J., Nojiri, H., Rachi, T., Tanigaki, K., Toyota, N. (2006). Ferromagnetic ordering in a new nickel polyborate $NiB_{12}O_{14}(OH)_{10}$. *Dalton Trans., 13*, 1597-1601.

Kaneta, T., Yamamoto, D., Imasaka, T. (2009). Postcolumn derivatization of proteins in capillary sieving electrophoresis/laser-induced fluorescence detection. *Electrophoresis, 30*, 3780-3785.

Kanie, Y., Enomoto, A., Goto, S., Kanie, O. (2008). Comparative RP-HPLC for rapid identification of glycopeptides and application in off-line LC-MALDI-MS analysis. *Carbohydr. Res., 343*, 758-768.

Kataoka, H., Okamoto, Y., Tsukahara, S., Fujiwara, T., Ito, K. (2008). Separate vaporisation of boric acid and inorganic boron from tungsten sample cuvette-tungsten boat furnace followed by the detection of boron species by inductively coupled plasma mass spectrometry and atomic emission spectrometry (ICP-MS and ICP-AES). *Anal. Chim. Acta, 610*, 179-185.

Katz, J.D., Lapointe, B.T., Dinsmore, C.J. (2009). Preparation of a stable trifluoroborate salt for the synthesis of 1-aryl-2,2-difluoro-enolethers and/or 2,2-difluoro-1-arylketones via palladium-mediated cross-coupling. *J. Org. Chem., 74*, 8866-8869.

Kawamoto, H., Saito, S., Saka, S. (2008). Inhibition of acid-catalyzed depolymerization of cellulose with boric acid in non-aqueous acidic media. *Carbohydr. Res., 343*, 249-255.

Kazarian, A.A., Smith, J.A., Hilder, E.F., Breadmore, M.C., Quirino, J.P., Suttil, J. (2010). Development of a novel fluorescent tag O-2-[aminoethyl] fluorescein for the electrophoretic separation of oligosaccharides. *Anal. Chim. Acta, 662*, 206-213.

Kelani, K.M. (2004). Selective potentiometric determination of zolpidem hemitartrate in tablets and biological fluids by using polymeric membrane electrodes. *J. AOAC Int., 87*, 1309-1318.

Kijewska, M., Kluczyk, A., Stefanowicz, P., Szewczuk, Z. (2009). Electrospray ionization mass spectrometric analysis of complexes between peptide-derived Amadori products and borate ions. *Rapid Commun. Mass Spectrom., 23*, 4038-4046.

Kim, D.H., Marbon, B.N., Faull, K.F., Eckhert, C.D. (2003). Esterification of borate with NAD^+ and NADH as studied by electrospray ionization mass spectrometry and ^{11}B NMR spectroscopy. *J. Mass Spectrom., 38*, 632-640.

Kim, D.H., Faull, K.F., Norris, A.J., Eckhert, C.D. (2004). Borate-nucleotide complex formation depends on charge and phosphorylation state. *J. Mass Spectrom., 39*, 743-751.

Kim, N.H., Kim, H.J., Kang, D.I., Jeong, K.W., Lee, J.K., Kim, Y., Oh, D.K. (2008). Conversion shift of D-fructose to D-psicose for enzyme-catalyzed epimerization by addition of borate. *Appl. Environ. Microbiol., 74*, 3008-3013.

Klaer, A., Syha, Y., Nasiri, H.R., Müller, T. (2009). Trisilyl-substituted vinyl cations. *Chemistry, 15*, 8414-8423.

Klimin, S.A., Fausti, D., Meetsma, A., Bezmaternykh, L.N., van Loosdrecht, P.H., Palstra, T.T. (2005). Evidence for differentiation in the iron-helicoidal chain in GdFe$_3$(BO$_3$)$_4$. *Acta Crystallogr. B., 61*, 481-485.

Kodama, S., Yamamoto, A., Matsunaga, A., Yanai, H. (2004b). Direct enantioseparation of catechin and epicatechin in tea drinks by 6-O-alpha-D-glucosyl-beta-cyclodextrin-modified micellar electrokinetic chromatography. *Electrophoresis, 25*, 2892-2898.

Kodama, S., Aizawa, S., Taga, A., Yamashita, T., Yamamoto, A. (2006). Chiral resolution of monosaccharides as 1-phenyl-3-methyl-5-pyrazolone derivatives by ligand-exchange CE using borate anion as a central ion of the chiral selector. *Electrophoresis, 27*, 4730-4734.

Kodama, S., Yamamoto, A., Iio, R., Sakamoto, K., Matsunaga, A., Hayakawa, K. (2004a). Chiral ligand exchange capillary electrophoresis using borate anion as a central ion. *Analyst, 129*, 1238-1242.

Kodama, S., Aizawa, S., Taga, A., Yamashita, T., Kemmei, T., Yamamoto, A., Hayakawa, K. (2007). Simultaneous chiral resolution of monosaccharides as 8-aminonaphthalene-1,3,6-trisulfonate derivatives by ligand-exchange CE using borate as a central ion of the chiral selector. *Electrophoresis, 28*, 3930-3933.

Kühn, K.D., Weber, C., Kreis, S., Holzgrabe, U. (2008). Evaluation of the stability of gentamicin in different antibiotic carriers using a validated MEKC method. *J. Pharm. Biomed. Anal., 48*, 612-618.

Kumar Malik, A., Faubel, W. (2000). Capillary electrophoretic determination of zinc dimethyldithiocarbamate (Ziram) and zinc ethylenebisdithiocarbamate (Zineb). *Talanta, 52*, 341-346.

Kumar, S., Kayastha, A.M. (2010). Inhibition studies of soybean (Glycine max) urease with heavy metals, sodium salts of mineral acids, boric acid, and boronic acids. *J. Enzyme Inhib. Med. Chem.*, doi: 10.3109/14756360903468155.

Kumar, J.S., Jonnalagadda, S.C., Mereddy, V.R. (2010a). An efficient boric acid mediated preparation of alpha-hydroxyamides. *Tetrahedron Lett.*, *51*, 779-782.

Kumar, M., Papish, E.T., Zeller, M., Hunter, A.D. (2010b). New alkylzinc complexes with bulky tris(triazolyl)borate ligands: surprising water stability and reactivity. *Dalton Trans.*, *39*, 59-61.

Kuo, C.Y., Wu, S.M. (2005). Micellar electrokinetic chromatography for simultaneous determination of six corticosteroids in commercial pharmaceuticals. *J. Sep. Sci.*, *28*, 144-148.

Kuriyama, M., Shimazawa, R., Enomoto, T., Shirai, R. (2008). Palladium-catalyzed 1,2-addition of potassium aryl- and alkenyltrifluoroborates to aldehydes using thioether-imidazolinium carbene ligands. *J. Org. Chem.*, *73*, 6939-6942.

Kusano, S., Ootani, A., Sakai, S., Igarashi, N., Takeguchi, A., Toyoda, H., Toida, T. (2007). HPLC determination of chondrosine in mouse blood plasma after intravenous or oral dose. *Biol. Pharm. Bull.*, *30*, 1365-1368.

Kusunose, T., Sakayanagi, N., Sekino, T., Ando, Y. (2008). Fabrication and characterization of aluminum nitride/boron nitride nanocomposites by carbothermal reduction and nitridation of aluminum borate powders. *J. Nanosci. Nanotechnol.*, *8*, 5846-5853.

Kwon, J.M., Kim, Y.H., Song, B.K., Yeom, S.H., Kim, B.S., Im, J.B. (2006). Novel immobilization of titanium dioxide (TiO_2) on the fluidizing carrier and its application to the degradation of azo-dye. *J. Hazard Mater.*, *134*, 230-236.

Jelen, F., Olejniczak, A.B., Kourilova, A., Lesnikowski, Z.J., Palecek, E. (2009). Electrochemical DNA detection based on the polyhedral boron cluster label. *Anal. Chem.*, *81*, 840-844.

Li, F., Cao, Q.E., Ding, Z. (2007). Separation and determination of three phenylpropanoids in the traditional Chinese medicine and its preparations by capillary electrophoresis. *J. Chromatogr. Sci.*, *45*, 354-359.

Li, S., Wang, J., Zhao, S. (2009b). Determination of terbutaline sulfate by capillary electrophoresis with chemiluminescence detection. *J. Chromatogr. B. Analyt. Technol. Biomed. Life Sci.*, *877*, 155-158.

Li, Q., Zhang, T., Lv, W. (2009a). A novel spectrophotometric method for the determination of aminophylline with boric acid in pharmaceutical and mixed serum samples. *Eur. J. Med. Chem.*, *44*, 1452-1456.

Li, J., Jiang, Y., Sun, T. Ren, S. (2008). Fast and simple method for assay of ciclopirox olamine by micellar electrokinetic capillary chromatography. *J. Pharm. Biomed. Anal.*, *47*, 929-933.

Li, F., Zhang, D., Lu, X., Wang, Y., Xiong, Z. (2004). Enatiomeric analysis of simendan by CE with beta-CD as chiral selector compared with CMPA-HPLC. *Biomed. Chromatogr.*, *18*, 866-871.

Li, C., Liu, J.X., Zhao, L., Di, D.L., Meng, M., Jiang, S.X. (2008). Capillary zone electrophoresis for separation and analysis of four diarylheptanoids and an alphatetralone derivative in the green walnut husks (*Juglans regia* L.). *J. Pharm. Biomed. Anal.*, *48*, 749-753.

Lim, B.C., Kim, H.J., Oh, D.K. (2007). High production of D-tagatose by the addition of boric acid. *Biotechnol. Prog.*, *23*, 824-828.

Lim, O., Suntornsuk, W., Suntornsuk, L. (2009). Capillary zone electrophoresis for enumeration of *Lactobacillus delbrueckii* subsp. *Bulgaricus* and *Streptococcus thermophilus* in yogurt. *J. Chromatogr. B Analyt. Technol. Biomed. Life Sci.*, *877*, 710-718.

Lin, S.L., Lin, C.E. (2004). Comparative studies on the enantioseparation of hydrobenzoin and structurally related compounds by capillary zone electrophoresis with sulfated beta-cyclodextrin as the chiral selector in the presence and absence of borate complexation. *J. Chromatogr. A.*, *1032*, 213-218.

Lin, C.E., Lin, S.L., Liao, W.S., Liu, Y.C. (2004a). Enantioseparation of benzoins and enantiomer migration reversal of hydrobenzoin in capillary zone electrophoresis with dual cyclodextrin systems and borate complexation. *J. Chromatogr. A.*, *1032*, 227-235.

Lin, X., Xue, L., Zhang, H., Zhu, C. (2005b). Determination of saikosaponins a, c, and d in Bupleurum Chinese DC from different areas by capillary zone electrophoresis. *Anal. Bioanal. Chem.*, *382*, 1610-1615.

Lin, C.E., Lin, S.L., Fang, I.J., Liao, W.S., Chen, C.C. (2004b). Enantioseparations of hydrobenzoin and structurally related compounds in capillary zone electrophoresis using heptakis-(2,3-dihydroxy-6-O-sulfo)-beta-cyclodextrin as chiral selector and enantiomer migration reversal of hydrobenzoin with a dual cyclodextrin system in the presence of borate complexation. *Electrophoresis*, *25*, 2786-2794.

Lin, C.E., Lin, S.L., Cheng, H.T., Fang, I.J., Kuo, C.M., Liu, Y.C. (2005a). Migration behavior and enantioseparation of hydrobenzoin and structurally related compounds in capillary zone electrophoresis with a dual cyclodextrin system consisting of heptakis-(2,3-dihydroxy-6-O-sulfo)-beta-cyclodextrin and beta-cyclodextrin. *Electrophoresis*, *26*, 4187-4196.

Liu, F.H., Jiang, Y. (2007). Room temperature ionic liquid as matrix medium for the determination of residual solvents in pharmaceuticals by static headspace gas chromatography. *J. Chromatogr. A.*, *1167*, 116-119.

Liu, J., Lou, Y.J. (2004). Determination of icariin and metabolites in rat serum by capillary zone electrophoresis: rat pharmacokinetic studies after administration of icariin. *J. Pharm. Biomed. Anal.*, *36*, 365-370.

Liu, C.Y., Chen, T.H., Misra, T.K. (2007). A macrocyclic polyamine as an anion receptor in the capillary electrochromatographic separation of carbohydrates. *J. Chromatogr. A.*, *1154*, 407-415.

Liu, J., Zhu, F., Liu, Z. (2009d). Borate complexation-assisted field-enhanced sample injection for online preconcentration of cis-diol-containing compounds in capillary electrophoresis. *Talanta*, *80*, 544-550.

Liu, H., Yea, X., Lia, W., Donga, Y., Wua, Z. (2009e). Comparison of boric acid adsorption by hybrid gels. *Desalin. Water Treat.*, *2*, 185-194.

Liu, X., Ye, S., Qiao, Y., Dong, G., Zhang, Q., Qiu, J. (2009a). Facile synthetic strategy for efficient and multi-color fluorescent BCNO nanocrystals. *Chem. Commun. (Camb.)*, *21*, 4073-4075.

Liu, P., He, W., Zhao, Y., Wang, P.A., Sun, X.L., Li, X.Y., Zhang, S.Y. (2008). Synthesis of chiral vicinal diols and analysis of them by capillary zone electrophoresis. *Chirality*, *20*, 75-83.

Liu, P., Sun, X., He, W., Jiang, R., Wang, P., Zhao, Y., Zhang, S. (2009c). Enantioselective separation of chiral vicinal diols in capillary electrophoresis using a mono-6(A)-aminoethylamino-beta-cyclodextrin as a chiral selector. *J. Sep. Sci.*, *32*, 125-134.

Liu, X., Huang, W., Fu, H., Yao, A., Wang, D., Pan, H., Lu, W.W., Jiang, X., Zhang, X. (2009b). Bioactive borosilicate glass scaffolds: in vitro degradation and bioactivity behaviors. *J. Mater. Sci. Mater. Med.*, *20*, 1237-1243.

Lockett, V., Sedev, R., Bassell, C., Ralston, J. (2008). Angle-resolved X-ray photoelectron spectroscopy of the surface of imidazolium ionic liquids. *Phys. Chem. Chem. Phys.*, *10*, 1330-1335.

London, R.E., Gabel, S.A. (2002). Formation of trypsin-borate-4-aminobutanol ternary complex. *Biochemistry*, *41*, 5963-5967.

Long, J.S., Silvester, D.S., Wildgoose, G.G., Surkus, A.E., Flechsig, G.U., Compton, R.G. (2008). Direct electrochemistry of horseradish peroxidase immobilized in a chitosan-[C_4mim] [BF_4] film: determination of electrode kinetic parameters. *Bioelectrochemistry*, *74*, 183-187.

Longinotti, M.P., Corti, H.R., Pablo, J.J. (2008). Secondary relaxations in supercooled and glassy sucrose-borate aqueous solutions. *Carbohydr. Res.*, *343*, 2650-2656.

Lu, C., Kostanski, L., Ketelson, H., Meadows, D., Pelton, R. (2005). Hydroxypropyl guar-borate interactions with tear film mucin and lysozyme. *Langmuir, 21*, 10032-10037.

Lu, W., Chen, Y., Zhang, Y., Ding, X., Chen, H., Liu, M. (2007). Microemulsion electrokinetic chromatography for the separation of arctiin and arctigenin in *Fructus arctii* and its herbal preparations. *J. Chromatogr. B. Analyt. Technol. Biomed. Life Sci.*, *860*, 127-133.

Luo, G.M., Tan, X.H., Xu, L.F., Yang, Y.Q., Yang, S.L. (2008). Study on fingerprints of Citrus aurantium from different places by capillary electrophoresis. *Zhongguo Zhong Yao Za Zhi, 33*, 2362-2364.

Luo, M., Lu, H., Ma, H., Zhao, L., Liu, X., Jiang, S. (2007). Separation and determination of flavonoids in *Lamiophlomis rotata* by capillary electrophoresis using borate as electrolyte. *J. Pharm. Biomed. Anal.*, *44*, 881-886.

Luykx, D.M., Cordewener, J.H., Ferranti, P., Frankhuizen, R., Bremer, M.G., Hooijerink, H., America, A.H. (2007). Identification of plant proteins in adulterated skimmed milk powder by high-performance liquid chromatography-mass spectrometry. *J. Chromatogr. A.*, *1164*, 189-197.

Lyhne-lversen, L., Hobley, T.J., Kaasgaard, S.G., Harris, P. (2006). Structure of *Bacillus halmapalus* alpha-amylase crystallized with and without the substrate analogue acarbose and maltose. *Acta Crystallogr. Sect. F. Struct. Biol. Cryst. Commun.*, *62*, 849-854.

Lyon, D.J., Spann, K.P. (1985). Assay of calcium borogluconate veterinary medicines for calcium gluconate, boric acid, phosphorous and magnesium by using inductively coupled plasma emission spectrometry. *J. Assoc. Off. Anal. Chem.*, *68*, 160-162.

Macholan, L., Skladal, P., Bohackova, I., Krejci, J. (1992). Amperometric glucose biosensor with extended concentration range utilizing complexation effect of borate. *Biosens. Bioelectron.*, *7*, 593-598.

Maeda, E., Hirano, K., Baba, Y., Nagata, H., Tabuchi, M. (2006). Conformational separation of monosaccharides of glycoproteins labeled with 2-aminoacrydone using microchip electrophoresis. *Electrophoresis, 27*, 2002-2010.

Maeda, E., Kataoka, M., Hino, M., Kajimoto, K., Kaji, N., Tokeshi, M., Kido, J., Shinohara, Y., Baba, Y. (2007). Determination of human blood glucose levels using microchip electrophoresis. *Electrophoresis, 28*, 2927-2933.

Maki, T., Ishihara, K., Yamamoto, H. (2005). N-alkyl-4-boronopyridinium-halides versus boric acid as catalysts for the esterification of alpha-hydroxycarboxylic acids. *Org. Lett., 27,* 5047-5050.

Malencik, D.A., Anderson, S.R. (1991). Fluorimetric characterization of dityrosine: complex formation with boric acid and borate ion. *Biochem. Biophys. Res. Commun., 178,* 60-67.

Mallampati, S., Leonard, S., De Vulder, S., Hoogmartens, J., Van Schepdael, A. (2005). Method development and validation for the analysis of didanosine using micellar electrokinetic capillary chromatography. *Electrophoresis, 26,* 4079-4088.

Malshe, K.M., Patil, P.T., Umbarkar, S.B., Dongare, M.K. (2004). Selective *C*-methylation of phenol with methanol over borate zirconia solid catalyst. *J. Mol. Catal. A: Chem., 212,* 337-344.

Manickum, T. (2008). Interferences by anti-TB drugs in a validated HPLC assay for urinary catecholamines and their successful removal. *J. Chromtogr. B. Analyt. Technol. Biomed. Life Sci., 873,* 124-128.

Mankad, N.P., Peters, J.C. (2008). Diazoalkanes react with a bis(phosphino)borate copper(i) source to generate $[Ph_2BP^{tBu}_2]Cu(\eta^1\text{-}N_2CR_2)$, $[Ph_2BP^{tBu}_2]Cu(CR_2)$, and $[Ph_2BP^{tBU}_2]Cu\text{-}N(CPh_2)(NCPh_2)$. *Chem. Commun. (Camb), 9,* 1061-1063.

Manoravi, P., Joseph, M., Sivakumar, N., Balasubramanian, R. (2005). Determination of isotopic ratio of boron in boric acid using laser mass spectrometry. *Anal. Sci., 21,* 1453-1455.

Marin-Zamora, M.E., Rojas-Melgarejo, F., Garcia-Canovas, F., Garcia-Ruiz, P.A. (2009). Production of o-diphenols by immobilized mushroom tyrosinase. *J. Biotechnol., 139,* 163-168.

Mary, S.S., Kirupavathy, S.S., Mythili, P., Srinivasan, P., Kanagasekaran, T., Gopalakrishnan, R. (2008). Studies on the growth, optical, electrical and spectral properties of potassium pentaborate (KB5) single crystals. *Spectrochim. Acta A Mol. Biomol. Spectrosc., 71,* 10-16.

Mason, P.E., Schildt, D.C., Strein, T.G. (2009). In-capillary determination of creatinine with electrophoretically mediated microanalysis: characterization of the effects of reagent zone and buffer conditions. *J. Chromatogr. A, 1216,* 154-158.

Matthews, D.A., Alden, R.A., Birktoft, J.J., Freer, S.T., Kraut, J. (1975). X-ray crystallographic study of boronic acid with subtilisin BPN' (Novo). A model for the catalytic transition state. *J. Biol. Chem., 250,* 7120-7126.

McMillen, C.D., Hu, J., VanDerveer, D., Kolis, J.W. (2009). Trigonal structures of $ABe_2BO_3F_2$ (A = Rb, Cs, Tl) crystals. *Acta Crystallogr. B*, 65, 445-449.

Meiland, M., Heinze, T., Guenther, W., Liebert, T. (2010). Studies on the boronation of methyl-beta-D-cellobioside-a cellulose model. *Carbohydr. Res.*, 345, 257-263.

Mendham, J., Denney, R.C., Barnes, J.D., Thomas, M.J.K. (2000). *Vogel's Textbook of Quantitative Chemical Analysis*, 6th ed., Delhi: Dorling Kindersley, pp. 384-385.

Menter, J.M., Abukhalaf, I.K., Patta, A.M., Silvestrov, N.A., Willis, I. (2007). Fluorescence of putative chromophores in Skh-1 and citrate-soluble calf skin collagens. *Photodermatol. Photoimmunol. Photomed.*, 23, 222-228.

Michalska, K., Pajchel, G., Tyski, S. (2009). Different sample stacking strategies for the determination of ertapenem and its impurities by micellar electrokinetic chromatography in pharmaceutical formulation. *J. Chromatogr. A.*, 1216, 2934-2942.

Min, X., Chai, L., Zhang, C., Takasaki, Y., Okura, T. (2008). Control of metal toxicity, effluent COD and regeneration of gel beads by immobilized sulfate-reducing bacteria. *Chemosphere*, 72, 1086-1091.

Mirza, M.A., Khuhawar, M.Y., Arain, R. (2008). Determination of uranium, iron, copper, and nickel in rock and water samples by MEKC. *J. Sep. Sci.*, 31, 3037-44.

Mishra, V., Nag, V.L., Tandon, R., Awasthi, S. (2010). Response surface methodology-based optimisation of agarose gel electrophoresis for screening and electropherotyping of rotavirus. *Appl. Biochem. Biotechnol.*, 160, 2322-2331.

Mohamed, A.A., Mubarak, A.T., Marstani, Z.M., Fawy, K.F. (2006). A novel kinetic determination of dissolved chromium species in natural and industrial waste water. *Talanta*, 70, 460-467.

Molander, G.A., Canturk, B. (2008). Preparation of potassium alkoxymethyltrifluoroborates and their crosscoupling with aryl chlorides. *Org. Lett.*, 10, 2135-2138.

Molander, G.A., Canturk, B. (2009). Organotrifluoroborates and monocoordinated palladium complexes as catalysts--a perfect combination for Suzuki-Miyaura coupling. *Angew. Chem. Int. Ed. Engl.*, 48, 9240-9261.

Molander, G.A., Cooper, D.J. (2008). Functionalization of organotrifluoroborates: reductive amination. *J. Org. Chem.*, 73, 3885-3891.

Molander, G.A., Ellis, N.M. (2008). Highly stereoselective synthesis of cis-alkenyl pinacolboronates and potassium cisalkenyltrifluoroborates via a hydroboration/ protodeboronation approach. *J. Org. Chem.*, *73*, 6841-6844.

Molander, G.A., Gormisky, P.E. (2008). Cross-coupling of cyclopropyl- and cyclobutyltrifluoroborates with aryl and heteroaryl chlorides. *J. Org. Chem.*, *73*, 7481-7485.

Molander, G.A., Jean-Gérard, L. (2009). Use of potassium beta-trifluoroborato amides in Suzuki-Miyaura cross-coupling reactions. *J. Org. Chem.*, *74*, 5446-5450.

Molander, G.A., Petrillo, D.E. (2008). Suzuki-Miyaura cross-coupling of potassium trifluoroboratohomoenolates. *Org. Lett.*, *10*, 1795-1798.

Molander, G.A., Sandrock, D.L. (2009). Potassium trifluoroborate salts as convenient, stable reagents for difficult alkyl transfers. *Curr. Opin. Drug Discov. Devel.*, *12*, 811-823.

Molander, G.A., Canturk, B., Kennedy, L.E. (2009a). Scope of the Suzuki-Miyaura cross-coupling reactions of potassium heteroaryltrifluoroborates. *J. Org. Chem.*, *74*, 973-980.

Molander, G.A., Febo-Ayala, W., Jean-Gérard, L. (2009b). Condensation reactions to form oxazolinesubstituted potassium organotrifluoroborates. *Org. Lett.*, *11*, 3830-3833.

Molander, G.A., Febo-Ayala, W., Ortega-Guerra, M. (2008b). Facile synthesis of highly functionalized ethyltrifluoroborates. *J. Org. Chem.*, *73*, 6000-6002.

Molander, G.A., Gormisky, P.E., Sandrock, D.L. (2008a). Scope of aminomethylations via Suzuki-Miyaura cross-coupling of organotrifluoroborates. *J. Org. Chem.*, *73*, 2052-2057.

Moore, J.A. (1997). An assessment of boric acid and borax using the IEHR evaluative process for assessing human developmental and reproductive toxicity of agents. Expert Scientific Committee. *Reprod. Toxicol.*, *11*, 123-160.

Morin, P., Villard, F., Drewx, M., Andre, P. (1993). Borate complexation of flavonoid-O-glycosides in capillary electrophoresis. II. Separation of flavonoid-3-O-glycosides differing in their sugar moiety. *J. Chromatogr.*, *628*, 161-169.

Murray, F.J. (1995). A human health risk assessment of boron (boric acid and borax) in drinking water. *Regul. Toxicol. Pharmacol.*, *22*, 221-230.

Musashi, M., Oi, T., Matsuo, M., Nomura, M. (2008). Column chromatographic boron isotope separation at 5 and 17 MPa with diluted boric acid solution. *J. Chromatogr. A., 1201*, 48-53.

Mutseneck, E.V., Bieller, S., Bolte, M., Lerner, H.W., Wagner, M. (2010). Fourth generation scorpionates: coordination behavior of a new class of conformationally flexible mixed-donor (pyrazol-1-yl) borates. *Inorg. Chem., 49*, 3540-3552.

Nakano, E., Iwatsuki, S., Inamo, M., Takagi, H.D., Ishihara, K. (2008). Optimization of conditions for the determination of boron by a ruthenium(II) complex having diol moiety: A mechanistic study. *Talanta, 74*, 533-538.

Navarre, L., Martinez, R., Genet, J.P., Darses, S. (2008). Access to enantioenriched alpha-amino esters via rhodium-catalyzed 1,4-addition/enantioselective protonation. *J. Am. Chem. Soc., 130*, 6159-6169.

Ogawa, S., Toyoda, M., Tonogal, Y., Ito, Y., Iwaida, M. (1979). Colorimetric determination of boric acid in prawns, shrimps and salted jelly fish by chelate extraction with 2-ethyl-1,3-hexanediol. *J. Assoc. Off. Anal. Chem., 62*, 610-614.

Oonuki, Y., Yoshida, Y., Uchiyama, Y., Asari, A. (2005). Application of fluorophore-assisted carbohydrate electrophoresis to analysis of disaccharides and oligosaccharides derived from glycosaminoglycans. *Anal. Biochem., 343*, 212-222.

Oudhoff, K.A., VanDamme, F.A., Mes, E.P., Schoenmakers, P.J., Kok, W.T. (2004). Characterization of glycerin-based polyols by capillary electrophoresis. *J. Chromatogr. A, 1046*, 263-269.

Ozbas, B., Rajagopal, K., Haines-Butterick, L., Schneider, J.P., Pochan, D.J. (2007). Reversible stiffening transition in beta-heparin hydrogels induced by ion complexation. *J. Phys. Chem. B, 111*, 13901-13908.

Öztürk, N., Kavak, D. (2005). Adsorption of boron from aqueous solutions using fly ash: Batch and column studies. *J. Hazard. Mater., B127*, 81-88.

Pacenti, M., Dugheri, S., Villanelli, F., Bartolucci, G., Calamai, L., Boccalon, P., Arcangeli, G., Vecchione, F., Alessi, P., Kikic, I., Cupelli, V. (2008). Determination of organic acids in urine by solid-phase microextraction and gas chromatography-ion trap tandem mass spectrometry previous 'in sample' derivatization with trimethyloxonium tetrafluoroborate. *Biomed. Chromatogr., 22*, 1155-63.

Palleschi, A., Coviello, T., Bocchinfuso, G., Alhaique, F. (2006). Investigation on a new scleroglucan/borax hydrogel: structure and drug release. *Int. J. Pharm., 322*, 13-21.

Pascali, J.P., Liotta, E., Gottardo, R., Bortolotti, F., Tagliaro, F. (2009). Rapid optimized separation of bromide in serum samples with capillary zone electrophoresis by using glycerol as additive to the background electrolyte. *J. Chromatogr. A.*, *1216*, 3349-3352.

Payan, E., Presle, N., Lapicque, F., Jouzeau, J.Y., Bordji, K., Qerther, S., Miralles, G., Mainard, D., Netter, P. (1998). Separation and quantification by ion-association capillary zone electrophoresis of unsaturated disaccharides units of chondroitin sulfates and oligosaccharides derived from hyaluronan. *Anal. Chem., 70*, 4780-4786.

Peak, D., Luther III, G.W., Sparks, D.L. (2003). ATR-FTIR spectroscopic studies of boric acid adsorption on hydrous ferric oxide. *Geochimica et Cosmochimica Acta, 67*, 2551-2560.

Peng, M., Wondraczek, L. (2009). Bi2+-doped strontium borates for white-light-emitting diodes. *Opt. Lett., 34*, 2885-2887.

Peng, Y.Y., Ye, J.N. (2006). Determination of isoflavones in red clover by capillary electrophoresis with electrochemical detection. *Fitoterapia, 77*, 171-178.

Penn, S.G., Hu, H., Brown, P.H., Lebrilla, C.B. (1997). Direct analysis of sugar alcohol borate complexes in plant extracts by matrix-assisted laser desorption / ionization Forrier transform mass spectrometry. *Anal. Chem., 69*, 2471-2477.

Periasamy, M., Kumar, N.S., Sivakumar, S., Rao, V.D., Ramanathan, C.R., Venkatraman, L. (2001). New methods of resolution and purification of racemic and diastereomeric amino alcohol derivatives using boric acid and chiral 1,1'-bis-2-naphthol. *J. Org. Chem., 66*, 3828-3833.

Petr, J., Vítková, K., Ranc, V., Znaleziona, J., Maier, V., Knob, R., Sevcík, J. (2008). Determination of some phenolic acids in *Majorana hortensis* by capillary electrophoresis with online electrokinetic preconcentration. *J. Agric. Food Chem., 56*, 3940-3944.

Petrásková, L., Charvátová, A., Prikrylová, V., Kristová, V., Rauvolfová, J., Martínková, L., Jiménez-Barbero, J., Aboitiz, N., Petrus, L., Kren, V. (2006). Preparative production and separation of 2-acetamido-2-deoxymannopyranoside-containing saccharides using borate-saturated polyolic exclusion gels. *J. Chromatogr. A., 1127*, 126-136.

Polat, H., Vengosh, A., Pankratov, I., Polat, M. (2004). A new methodology for removal of boron from water by coal and fly ash. *Desalination, 164*, 173-188.

Pospisilova, M., Polasek, M., Jokl, V. (1998). Separation and determination of sorbitol and xylitol in multicomponent pharmaceutical formulations by capillary isotachophoresis. *J. Pharm. Biomed. Anal., 17*, 387-392.

Puig, P., Borrull, F., Calull, M., Aguilar, C. (2005). Sample stacking for the analysis of eight penicillin antibiotics by micellar electrokinetic capillary chromatography. *Electrophoresis, 26*, 954-961.

Punzet, M., Ferreira, F., Briza, P., van Ree, R., Malissa, H. Jr., Stutz, H. (2006). Profiling preparations of recombinant birch pollen allergen Bet v 1a with capillary zone electrophoresis in pentamine modified fused-silica capillaries. *J. Chromatogr. B. Analyt. Technol. Biomed. Life Sci., 839*, 19-29.

Qi, L., Yang, G. (2009a). On-column labeling technique and chiral ligandexchange CE with zinc(II)-L-arginine complex as a chiral selector for assay of dansylated D,L amino acids. *Electrophoresis, 30*, 2882-2889.

Qi, L., Yang, G. (2009b). Enantioseparation of dansyl amino acids by ligand-exchange capillary electrophoresis with zinc(II)-L-phenylalaninamide complex. *J. Sep. Sci., 32*, 3209-3214.

Qi, S., Cui, S., Chen, X., Hu, Z. (2004). Rapid and sensitive determination of anthraquinones in Chinese herb using 1-butyl-3-methylimidazolium-based ionic liquid with beta-cyclodextrin as modifier in capillary zone electrophoresis. *J. Chromatogr. A., 1059*, 191-198.

Qi, L., Han, Y., Zuo, M., Chen, Y. (2007). Chiral CE of aromatic amino acids by ligandexchange with zinc(II)-L-lysine complex. *Electrophoresis, 28*, 2629-2634.

Qi, L., Chen, Y., Xie, M., Guo, Z., Wang, X. (2008). Separation of dansylated amino acid enantiomers by chiral ligand-exchange CE with a zinc(II) L-arginine complex as the selectingsystem. *Electrophoresis, 29*, 4277-4283.

Qi, L., Cui, K., Qiao, J., Yang, G., Chen, Y. (2009). Use of MEKC for the analysis of reactant and product of Baylis-Hillman reaction. *J. Sep. Sci., 32*, 1480-1486.

Qi, S., Li, Y., Deng, Y., Cheng, Y., Chen, X., Hu, Z. (2006). Simultaneous determination of bioactive flavone derivatives in Chinese herb extraction by capillary electrophoresis used different electrolyte systems--borate and ionic liquids. *J. Chromatogr. A., 1109*, 300-306.

Qiu, W., Zeng, X. (2008). Conductive polymer as a controlled microenvironment for the potentiometric highthroughput evaluation of ionic liquid cell toxicity. *Anal. Bioanal. Chem., 392*, 203-213.

Ramdoss, K., Amma, B.G., Umashankar, V., Rangaswamy, R. (1997). Cold dissolution method for the determination of uranium in various geological materials at trace levels by laser fluorimetry. *Talanta*, *44*, 1095-1098.

Ramos, A., Lough, A.J., Stephan, D.W. (2009). Activation of H_2 by frustrated Lewis pairs derived from mono- and bisphosphinoferrocenes and $B(C_6F_5)_3$. *Chem. Commun. (Camb.)*, *9*, 1118-1120.

Reshak, A.H., Chen, X., Auluck, S., Kityk, I.V. (2008). X-ray diffraction and optical properties of a noncentrosymmetric borate $CaBiGaB_2O_7$. *J. Chem. Phys.*, *28*, 129, 204111.

Reshak, A.H., Auluck, S., Kityk, I.V., Chen, X. (2009a). X-ray diffraction, X-ray photoelectron spectra, crystal structure, and optical properties of centrosymmetric strontium borate $Sr_2B_{16}O_{26}$. *J. Phys. Chem. B.*, *113*, 9161-9167.

Reshak, A.H., Kityk, I.V., Auluck, S., Chen, X. (2009b). X-ray diffraction, crystal structure, and spectral features of the optical susceptibilities of single crystals of the ternary borate oxide lead bismuth tetraoxide, $PbBiBO_4$. *J. Phys. Chem. B.*, *113*, 6640-6646.

Rey, S., Davies, D.M. (2006). Photochemistry of peroxoborates: borate inhibition of the photodecomposition of hydrogen peroxide. *Chemistry*, *12*, 9284-9288.

Richold, M. (1998). Boron exposure from consumer products. *Biol. Trace Elem. Res.*, *66*, 121-129.

Robinson-Fuentes, V.A., Jaime-Sánchez, J.L., García-Aguilar, L., Gómez-Peralta, M., Vázquez-Garcidueñas, M.S., Vázquez-Marrufo, G. (2008). Determination of alpha- and beta-amanitin in clinical urine samples by capillary zone electrophoresis. *J. Pharm. Biomed. Anal.*, *47*, 913-917.

Rogers, H.R., van den Berg, C.M. (1988). Determination of borate ion-pair stability constants by potentiometry and non-approximative linearization of titration data. *Talanta*, *35*, 271-275.

Rohr, U., Meckea, L., Strubel, C. (2004). A methodology for the determination of reductive sulphur in optical and technical glass. *Talanta*, *63*, 933-939.

Rong, C., Yu, Z., Wang, Q., Zheng, S.T., Pan, C.Y., Deng, F., Yang, G.Y. (2009). Aluminoborates with open frameworks: syntheses, structures, and properties. *Inorg. Chem.*, *48*, 3650-3659.

Ruth, K., Tullmann, S., Vitze, H., Bolte, M., Lerner, H.W., Holthausen, M.C., Wagner, M. (2008). Copper complexes of mono- and ditopic [(methylthio)methyl]borates: missing links and linked systems En route to copper enzyme models. *Chemistry*, *14*, 6754-6770.

Sakurai, F., Kondo, K., Aoyama, T. (2009). Interesting nickel-catalyzed 1,2-addition to alpha,beta-unsaturated aldehydes with arylborates. *Chem. Pharm. Bull.*, *57*, 511-512.

Salentine, C.G. (1983). High field ^{11}B NMR of alkali borates. Aqueous polyborate equilibria. *Inorg. Chem.*, *22*, 3920-3924.

Sanchez-Ramos, S., Medina-Hernandez, M.J., Sagrado, S. (1998). Flow injection spectrophotometric determination of boron in ceramic materials. *Talanta*, *45*, 835-842.

Sano, A., Nakamura, H. (2007). Evaluation of titanium and titanium oxides as chemo-affinity sorbents for the selective enrichment of organic phosphates. *Anal. Sci.*, *23*, 1285-1289.

Sarkar, A., Alamelu, D., Aggarwal, S.K. (2009). Laser-induced breakdown spectroscopy for determination of uranium in thorium-uranium mixed oxide fuel materials. *Talanta*, *78*, 800-804.

Sathe, S.K., Venkatachalam, M., Sharma, G.M., Kshirsagar, H.H., Teuber, S.S., Roux, K.H. (2009). Solubilization and electrophoretic characterization of select edible nut seed proteins. *J. Agric. Food Chem.*, *57*, 7846-7856.

Saxena, R., Verma, R.M. (1983). Iodometric microdetermination of boric acid and borax separately or in a mixture. *Talanta*, *30*, 365-367.

Schreiner, C., Amereller, M., Gores, H.J. (2009). Chloride-free method to synthesise new ionic liquids with mixed borate anions. *Chemistry*, *15*, 2270-2272.

Schubert, D.M., Visi, M.Z., Knobler, C.B. (2008b). Crystalline alcoholamine borates and the triborate monoanion. *Inorg. Chem.*, *47*, 2017-2023.

Schubert, D.M., Visi, M.Z., Khan, S., Knobler, C.B. (2008a). Synthesis and structure of a new heptaborate oxoanion isomer: $B_7O_9(OH)_5^{2-}$. *Inorg. Chem.*, *47*, 4740-4745.

Sciarra, J.J., Monte Bovi, A.J. (1962). Study of the boric acid-glycerin complex. II. Formation of the complex at elevated temperatures. *J. Pharm. Sci.*, *51*, 238-242.

Sekar, R., Azhaguvel, S. (2005). Simultaneous determination of HIV-protease inhibitors lamivudine and zidovudine in pharmaceutical formulations by micellar electrokinetic chromatography. *J. Pharm. Biomed. Anal.*, *39*, 653-660.

Selander, N., Kipke, A., Sebelius, S., Szabo, K.J. (2007). Petasis Borono-Mannich reaction and allylation of carbonyl compounds via transient allyl boronates generated by palladium-catalyzed substitution of allyl alcohols.

An efficient one-pot route to stereodefined alpha-amino acids and homoallyl alcohols. *J. Am. Chem. Soc.*, *129*, 13723-13731.

Serrano, J.M., Silva, M. (2006). Trace analysis of aminoglycoside antibiotics in bovine milk by MEKC with LIF detection. *Electrophoresis*, *27*, 4703-4710.

Shuya, C., Shengda, Q., Xingguo, C., Zhide, H. (2004). Identification and determination of effective components in *Euphrasia regelii* by capillary zone electrophoresis. *Biomed. Chromatogr.*, *18*, 857-861.

Smith, K.W., Johnson, S.L. (1976). Borate inhibition of yeast dehydrogenase. *Biochemistry, 15*, 560-564.

Smith, L.M., Maher, A.D., Want, E.J., Elliott, P., Stamler, J., Hawkes, G.E., Holmes, E., Lindon, J.C., Nicholson, J.K. (2009). Large-scale human metabolic phenotyping and molecular epidemiological studies via ^{1}H NMR spectroscopy of urine: investigation of borate preservation. *Anal. Chem.*, *81*, 4847-4856.

Solangi, A.R., Bhanger, M.I., Memon, S.Q., Khuhawar, M.Y., Mallah, A. (2009). A capillary zone electrophoretic method for simultaneous determination of seven drugs in pharmaceuticals and in human urine. *J. AOAC Int.*, *92*, 1382-1389.

Song, Y., Cooks, R.G. (2007). Reactive desorption electrospray ionization for selective detection of the hydrolysis products of phosphonate esters. *J. Mass Spectrom.*, *42*, 1086-1092.

Sonnenschein, M.F., Redwine, O.D., Wendt, B.L., Kastl, P.E. (2009). Colloidal encapsulation of hydrolytically and oxidatively unstable organoborane catalysts and their use in waterborne acrylic polymerization. *Langmuir*, *25*, 12488-12494.

Speck, A.J., Odink, J., Schrijver, J., Schreurs, W.H. (1983). High-performance liquid chromatographic determination of urinary free catecholamines with electrochemical detection after prepurification on immobilized boric acid. *Clin. Chim. Acta, 28,* 103-113.

Strutz, K., Stellwagen, N.C. (1998). Do DNA gel electrophoretic mobilities extrapolate to the free-solution mobility of DNA at zero gel concentration? *Electrophoresis*, *19*, 635-642.

Subirats, X., Bosch, E., Rosacs, M. (2009). Retention of ionisable compounds on high-performance liquid chromatography XVIII: pH variation in mobile phases containing formic acid, piperazine, tris, boric acid or carbonate as buffering systems and acetonitrile as organic modifier. *J. Chromatogr. A.*, *1216*, 2491-2498.

Sun, G., Song, W., Lin, T. (2008). Selection of back-ground electrolyte in capillary zone electrophoresis by triangle and tetrahedron optimization methods. *Se Pu*, *26*, 232-236.

Süslü, I., Altinöz, S. (2005). Electrochemical characteristics of zafirlukast and its determination in pharmaceutical formulations by voltammetric methods. *J. Pharm. Biomed. Anal.*, *39*, 535-542.

Süslü, I., Demircan, S., Altinöz, S., Kir, S. (2007). Optimisation, validation and application of a capillary electrophoretic method for the determination of zafirlukast in pharmaceutical formulations. *J. Pharm. Biomed. Anal.*, *44*, 16-22.

Taler, G., Eliav, U., Navon, G. (1999). Detection and characterization of boric acid and borate ion binding to cytochrome *c* using multiple quantum filtered NMR. *J. Magn. Reson.*, *141*, 228-238.

Tamat, S.R., Moore, D.E., Allen, B.J. (1989). Determination of the concentration of complex boronated compounds in biological tissues by inductively coupled plasma atomic emission spectrometry. *Pigment Cell Res.*, *2*, 281-285.

Tamesue, S., Numata, M., Kaneko, K., James, T.D., Shinkai, S., (2008). Hierarchical carbon nanotube assemblies created by sugar-boric or boronic acid interactions. *Chem. Commun. (Camb.)*, *37*, 4478-4480.

Tanaka, K., Fujimori, Y., Saikawa, Y., Nakata, M. (2008). Diastereoselective synthesis of useful building blocks by crotylation of beta-branched alphamethylaldehydes with potassium crotyltrifluoroborates. *J. Org. Chem.*, *73*, 6292-6298.

Tang, Y., Wu, M. (2005). A quick method for the simultaneous determination of ascorbic acid and sorbic acid in fruit juices by capillary zone electrophoresis. *Talanta*, *65*, 794-798.

Tang, X.H., Wei, C.H., Liang, J.R., Wang, B.G. (2006). Preparation and photocatalytic activity of boron doped CeO_2/TiO_2 mixed oxides. *Huan Jing Ke Xue*, *27*, 1329-1333.

Tao, X., Li, X. (2008). Catalyst-free synthesis, structural, and mechanical characterization of twinned $Mg_2B_2O_5$ nanowires. *Nano Lett.*, *8*, 505-510.

Tenorio-López, F.A., Zarco-Olvera, G., Sánchez-Mendoza, A., Rosas-Peralta, M., Pastelín-Hernández, G., del Valle-Mondragón, L. (2010). Simultaneous determination of angiotensins II and 1-7 by capillary zone electrophoresis in plasma and urine from hypertensive rats. *Talanta*, *80*, 1702-1712.

Tian, K., Qi, S., Cheng, Y., Chen, X., Hu, Z. (2005). Separation and determination of lignans from seeds of Schisandra species by micellar electrokinetic capillary chromatography using ionic liquid as modifier. *J. Chromatogr. A.*, *1078*, 181-187.

Tian, Y., Feng, R., Liao, L., Liu, H., Chen, H., Zeng, Z. (2008). Dynamically coated silica monolith with ionic liquids for capillary electrochromatography. *Electrophoresis*, *29*, 3153-3159.

Transue, T.R., Gabel, S.A., London, R.E. (2006). NMR and crystallographic characterization of adventitious borate binding by trypsin. *Bioconjug. Chem.*, *17*, 300-308.

Tseng, H.M., Gattolin, S., Pritchard, J., Newbury, H.J., Barrett, D.A. (2009). Analysis of mono-, di- and oligosaccharides by CE using a two-stage derivatization method and LIF detection. *Electrophoresis*, *30*, 1399-1405.

Tsuboi, T., Hirano, Y., Shibata, Y., Motomizu, S. (2002). Sensitivity improvement of ammonia determination based on flow-injection indophenol spectrophotometry with manganese (II) ion as a catalyst and analysis of exhaust gas of thermal power plant. *Anal Sci.*, *18*, 1141-1144.

Tulinsky, A., Blevins, R.A. (1987). Structure of a tetrahedral transition state complex of α-chymotrypsin dimer at 1.8 Å resolution. *J. Biol. Chem.*, *262*, 7737-7743.

United States Pharmacopeia 30 / National Formulary 25 (2007). Rockville, MD: United States Pharmacopeial Convention, Electronic Version.

Van Duin, M., Peters, J.A., Keiboom, A.P.G., Van Bekkum, H. (1987). Studies on boron esters, part 5. The system glucarate-borate-calcium (II) as studied by ^{1}H, ^{11}B and ^{13}C nuclear magnetic resonance spectroscopy. *J. Chem. Soc. Perkin Trans.*, *2*, 473-478.

Van Staden, J.K., Tsanwani, M.M. (2002). Determination of boron as boric acid in eye lotions using a sequential injection system. *Talanta*, *58*, 1103-1108.

Velchuri, R., Vijaya Kumar, B., Rama Devi, V., Ravi Kumar, K., Prasad, G., Vithal, M. (2009). Low temperature preparation and characterization of $In_{1-x}Ln_xBO_3$ ($x = 0.0$ and 0.05; Ln = Gd, Eu, Dy and Sm): ESR of $In_{0.95}Gd_{0.05}BO_3$ and emission of $In_{0.95}Eu_{0.05}BO_3$. *Spectrochim. Acta A Mol. Biomol. Spectrosc.*, *74*, 726-730.

Vieira, E.F., Cestari, A.R., Airoldi, C., Loh, W. (2008). Polysaccharide-based hydrogels: preparation, characterization, and drug interaction behaviour. *Biomacromolecules*, *9*, 1195-1199.

Villasuso, A.L., Racagni, G.E., Machado, E.E. (2008). Phosphatidylinositol kinases as regulators of GA-stimulated alpha-amylase secretion in barley (Hordeum vulgare). *Physiol. Plant*, *133*, 157-166.

Volpi, N. (2009). Capillary electrophoresis determination of glucosamine in nutraceutical formulations after labeling with anthranilic acid and UV detection. *J. Pharm. Biomed. Anal.*, *49*, 868-871.

Wagner, C.C., Ferraresi Curotto, V., Pis Diez, R., Baran, E.J. (2008). Experimental and theoretical studies of calcium fructoborate. *Biol. Trace Elem. Res.*, *122*, 64-72.

Wang, C., Armstrong, D.W., Risley, D.S. (2007). Empirical observations and mechanistic insights on the first boron-containing chiral selector for LC and supercritical fluid chromatography. *Anal. Chem.*, *79*, 8125-8135.

Wang, D.H., Mei, T.S., Yu, J.Q. (2008). Versatile Pd(II)-catalyzed C-H activation/arylaryl coupling of benzoic and phenyl acetic acids. *J. Am. Chem. Soc.*, *130*, 17676-17677.

Wang, X.H., Zheng, S.L., Xu, H.B., Zhang, Y. (2009a). Analysis of metallic elements in refractory tantalum-niobium slag by ICP-AES. *Guang Pu Xue Yu Guang Pu Fen Xi*, *29*, 805-808.

Wang, S., Alekseev, E.V., Stritzinger, J.T., Depmeier, W., Albrecht-Schmitt, T.E. (2010). How are centrosymmetric and noncentrosymmetric structures achieved in uranyl borates? *Inorg. Chem.*, *49*, 2948-2953.

Wang, W.G., Wang, H.Y., Si, G., Tung, C.H., Wu, L.Z. (2009b). Fluorophenyl-substituted Fe-only hydrogenases active site ADT models: different electrocatalytic process for proton reduction in HOAc and HBF_4/Et_2O. *Dalton Trans.*, *15*, 2712-2720.

Wang, Z., Tang, Z., Gu, Z., Hu, Z., Ma, S., Kang, J. (2005). Enantioseparation of chiral allenic acids by micellar electrokinetic chromatography with cyclodextrins as chiral selector. *Electrophoresis*, *26*, 1001-1006.

Waseem, A., Yaqoob, M., Nabi, A. (2004). Determination of iron in blood serum using flow injection with luminol chemiluminescence detection. *Luminescence*, *19*, 333-338.

Watson, D.G. (2005). *Pharmaceutical Analysis*, 2nd ed., London: Elsevier Churchill Livingstone, p. 369.

Wei, W., Yin, X.B., He, X.W. (2008). pH-mediated dual-cloud point extraction as a preconcentration and clean-up technique for capillary electrophoresis determination of phenol and m-nitrophenol. *J. Chromatogr. A.*, *1202*, 212-215.

Wester, R.C., Hartway, T., Maibacch, H.I., Schell, M.J., Northington, D.J., Culver, B.D., Strong, P.L. (1998a). In vitro percutaneous absorption of boron as boric acid, borax, and disodium octaborate tetrahydrate in human skin: a summary. *Biol. Trace Elem. Res.*, *66*, 111-120.

Wester, R.C., Hui, X., Maibach, H.I., Bell, K., Schell, M.J., Northington, D.J., Strong, P., Culver, B.D. (1998b). In vivo percutaneous absorption of boric acid, borax, and disodium octaborate tetrahydrate in humans: a summary. *Biol. Trace Elem. Res., 66,* 101-109.

Wester, R.C., Hui, X., Hartway, T., Maibach, H.I., Bell, K., Schell, M.J., Northington, D.J., Strong, E., Culver, B.D. (1998c). In vitro percutaneous absorption of boric acid, borax, and disodium octaborate tetrahydrate in humans compared to in vitro absorption in human skin from infinite and finite doses. *Toxicol. Sci., 45,* 42-51.

Whitaker, G., Lillquist, A., Pasas, S.A., O'Connor, R., Regan, F., Lunte, C.E., Smyth, M.R. (2008). CE-LIF method for the separation of anthracyclines: application to protein binding analysis in plasma using ultrafiltration. *J. Sep. Sci., 31,* 1828-1833.

Wu, J.J., Li, N., Li, K.A., Liu, F. (2008a). J-aggregates of disprotonated tetrakis(4-sulfonatophenyl)porphyrin induced by ionic liquid 1-butyl-3-methylimidazolium tetrafluoroborate. *J. Phys. Chem. B, 112,* 8134-8138.

Wu, L., Roth, G., Sparta, K., Chen, X. (2008c). The new pentaborate $Na_3SrB_5O_{10}$. *Acta Crystallogr. C, 64,* 53-56.

Wu, X., Wei, W., Su, Q., Xu, L., Chen, G. (2008b). Simultaneous separation of basic and acidic proteins using 1-butyl-3-methylimidazolium-based ion liquid as dynamic coating and background electrolyte in capillary electrophoresis. *Electrophoresis, 29,* 2356-2362.

Wu, X., Zhao, B., Wu, P., Zhang, H., Cai, C. (2009). Effects of ionic liquids on enzymatic catalysis of the glucose oxidase toward the oxidation of glucose. *J. Phys. Chem. B, 113,* 13365-13373.

Xiao, Q., Hu, B., He, M. (2008). Speciation of butyltin compounds in environmental and biological samples using headspace single drop microextraction coupled with gas chromatography-inductively coupled plasma mass spectrometry. *J. Chromatogr. A, 1211,* 135-141.

Xiaoping, L., Shiyang, G., Shuping, X. (2004). Investigations of kinetics and mechanism of chloropinnoite in boric acid aqueous solution at 303 K by Raman spectroscopy. *Spectrochim. Acta A Mol. Biomol. Spectrosc., 60,* 2725-2728.

Xie, T., Liu, Q., Shi, Y., Liu, Q. (2006). Simultaneous determination of positional isomers of benzenediols by capillary zone electrophoresis with square wave amperometric detection. *J. Chromatogr. A, 1109,* 317-321.

Xie, Y.N., Wang, S.F., Zhang, Z.L., Pang, D.W. (2008). Interaction between room temperature ionic liquid [bmim]BF_4 and DNA investigated by electrochemical micromethod. *J. Phys. Chem. B, 112,* 9864-9868.

Xu, F., Zhang, Z., Jiao, H., Tian, Y., Zhang, B., Chen, Y. (2006). Quantification of fudosteine in human plasma by high-performance liquid chromatography-electrospray ionization mass spectrometry employing precolumn derivatization with 9-fluorenylmethyl chloroformate. *J. Mass Spectrom.*, *41*, 685-692.

Yang, Y., Shi, X., Zhao, R. (1999). Flame retardancy behavior of zinc borate. *J. Fire Sci.*, *17*, 355-361.

Yang, R., Wang, Y., Zhou, D. (2007). Novel hydroxyethylcellulose-graft-poly acrylamide copolymer for separation of doublestranded DNA fragments by CE. *Electrophoresis*, *28*, 3223-3231.

Yang, X., Zhao, Y., Ruan, Y., Yang, Y. (2008b). Development and application of a capillary electrophoretic method for the composition analysis of a typical heteropolysaccharide from *Codonopsis pilosula* NANNF. *Biol. Pharm. Bull.*, *31*, 1860-1865.

Yang, C.H., Yang, G.F., Pan, Y.X., Zhang, Q.Y. (2009a). Synthesis and spectroscopic properties of $GdAl_3(BO_3)_4$ polycrystals codoped with Yb^{3+} and Eu^{3+}. *J. Fluoresc.*, *19*, 105-109.

Yang, T., Li, G., You, L., Ju, J., Liao, F., Lin, J. (2005). $MCuB_7O_{12}.nH_2O$ (M = Na, K): a new copper borate with 14-ring channels. *Chem. Commun. (Camb.)*, *33*, 4225-4227.

Yang, T., Sun, J., Eriksson, L., Li, G., Zou, X., Liao, F., Lin, J. (2008a). $Na_5[MB_{24}O_{34}(OH)_{12}].nH_2O$ (M = Cr^{3+}, Al^{3+}): unprecedented spherelike polyborate clusters from boric acid flux synthesis. *Inorg. Chem.*, *47*, 3228-3233.

Yang, T., Sun, J., Li, G., Wang, Y., Christensen, J., He, Z., Christensen, K.E., Zou, X., Liao, F., Lin, J. (2009b). $Fe_5O_5[B_6O_{10}OH_3].nH_2O$: Wave-Layered Iron Borate and Frustrated Antiferromagnetism. *Inorg. Chem.*, *48*, 11209-11214.

Yazicigil, Z., Oztekin, Y. (2006). Boron removal by electrodialysis with anion-exchange membranes. *Desalination*, *190*, 71-78.

Yilmaz, I., Kabay, N., Brjyak, M., Yüksel, M., Wolska, J., Koltuniewicz, A. (2006). A submerged membrane-ion-exchange hybrid process for boron removal. *Desalination*, *198*, 310-315.

Yohe, H.C. (1994). Removal of borate from tritiated gangliosides via the mannitoborate complex. *J. Lipid Res.*, *35*, 2100-2102.

Yoshida, M., Watabiki, T., Ishida, N. (1989b). Spectrophotometric determination of boric acid by the curcumin method. *Nihon Hoigaku Zasshi*, *43*, 490-496.

Yoshida, M., Watabiki, T., Tokiyasu, T., Ishida, N. (1989a). Determination of boric acid in biological materials by curcuma paper. *Nihon Hoigaku Zasshi, 43*, 497-501.

Yoshida, M., Tokiyasu, T., Watabiki, T., Ueda, M., Ishida, N. (1991). Study on the histochemical staining of boric acid. *Nihon Hoigaku Zasshi, 45*, 416-422.

Yoshida, J.L., Yoshino, K., Matsunaga, T., Higa, S., Suzuki, T., Yamamura, Y. (1980). An improved method for the determination of plasma norepinephrine: isolation by boric acid gel and assay by selected ion monitoring. *Biomed. Mass spectrom., 7*, 396-398.

Yoshinari, T., Forbes, R.T., York, P., Kawashima, Y. (2003). Crystallization of amorphous mannitol is retarded using boric acid. *Int. J. Pharm., 258*, 109-120.

You, J., Sheng, X., Ding, C., Sun, Z., Suo, Y., Wang, H., Li, Y. (2008). Detection of carbohydrates using new labeling reagent 1-(2-naphthyl)-3-methyl-5-pyrazolone by capillary zone electrophoresis with absorbance (UV). *Anal. Chim. Acta., 609*, 66-75.

Yu, S., Lindeman, S., Tran, C.D. (2008a). Chiral ionic liquids: synthesis, properties, and enantiomeric recognition. *J. Org. Chem., 73*, 2576-2591.

Yu, X.Q., Yamamoto, Y., Miyaura, N. (2008b). Aryl triolborates: novel reagent for coppercatalyzed N arylation of amines, anilines, and imidazoles. *Chem. Asian J., 3*, 1517-1522.

Zacharis, C.K., Tzanavaras, P.D., Notou, M., Zotou, A., Themelis, D.G. (2009). Separation and determination of nimesulide related substances for quality control purposes by micellar electrokinetic chromatography. *J. Pharm. Biomed. Anal., 49*, 201-206.

Zeng, H.L., Shen, H., Nakagama, T., Uchiyama, K. (2007). Property of ionic liquid in electrophoresis and its application in chiral separation on microchips. *Electrophoresis, 28*, 4590-4596.

Zhang, N., Zhang, X., Zhao, Y. (2004). Voltammetric study of the interaction of ciprofloxacin-copper with nucleic acids and the determination of nucleic acids. *Talanta, 62*, 1041-1045.

Zhang, N., Zhang, H.S., Wang, H. (2009b). Separation of free amino acids and catecholamines in human plasma and rabbit vitreous samples using a new fluorogenic reagent 3-(4-bromobenzoyl)-2-quinolinecarboxaldehyde with CE-LIF detection. *Electrophoresis, 30*, 2258-2265.

Zhang, Y., Chen, J., Zhao, L., Shi, Y.P. (2007b). Separation and determination of isoflavones in red clover by micellar electrokinetic capillary chromatography. *Biomed. Chromatogr., 21*, 987-992.

Zhang, J.P., Sun, Y., Yang, G., Li, Z.H. (2007a). Synthesis, characterization and study on vibration spectra of potassium triborate. *Guang Pu Xue Yu Guang Pu Fen Xi*, *27*, 1351-1354.

Zhang, J., Cui, H., Xu, L., Zhang, L., Chen, G. (2009a). Analysis of aliphatic amines using head-column field-enhanced sample stacking in MEKC with LIF detection. *Electrophoresis*, *30*, 674-681.

Zhang, F., Jiao, Z.W., Shen, D.Z., Shen, G.Q., Wang, X.Q. (2010). $CdZn_2KB_2O_6F$, a new fluoride borate crystal. *Acta Crystallogr. C*, *66*(Pt 1), i1-i3.

Zhang, N., Wang, H., Zhang, Z.X., Deng, Y.H., Zhang, H.S. (2008b). Sensitive determination of biogenic amines by capillary electrophoresis with a new fluorogenic reagent 3-(4-fluorobenzoyl)-2-quinolinecarboxaldehyde. *Talanta*, *76*, 791-797.

Zhang, Y., Desai, A., Lu, Z., Hu, G., Ding, Z., Wulff, W.D. (2008a). Catalytic asymmetric aziridination with borate catalysts derived from VANOL and VAPOL legends: scope and mechanistic studies. *Chemistry*, *14*, 3785-3803.

Zhang, S., Fei, Y., Frantz, E., Snyder, D.W., Chai, B.H., Shrout, T.R. (2008c). High-temperature piezoelectric single crystal $ReCa_4O(BO_3)_3$ for sensor applications. *IEEE Trans Ultrason. Ferroelectr. Freq. Control*, *55*, 2703-2708.

Zhang, H., Tian, K., Tang, J., Qi, S., Chen, H., Chen, X., Hu, Z. (2006). Analysis of baicalein, baicalin and wogonin in *Scutellariae radix* and its preparation by microemulsion electrokinetic chromatography with 1-butyl-3-methylimizolium tetrafluoborate ionic liquid as additive. *J. Chromatogr. A*, *1129*, 304-307.

Zhao, X., Wang, Y., Sun, Y. (2007). Simultaneous determination of four bioactive constituents in Liuwei Dihuang Pills by micellar electrokinetic chromatography. *J. Pharm. Biomed. Anal.*, *44*, 1183-1186.

Zhao, J., Zhang, Y., Cheng, K. (2008b). Palladium-catalyzed direct C-2 arylation of indoles with potassium aryltrifluoroborate salts. *J. Org. Chem.*, *73*, 7428-7431.

Zhao, S., Pan, Z., Chen, X., Hu, Z. (2004). Analysis of the aconitine alkaloids in traditional Chinese medicines by nonaqueous capillary electrophoresis using a new recording mode. *Biomed. Chromatogr.*, *18*, 381-387.

Zhao, S., Wang, J., Ye, F., Liu, Y.M. (2008a). Determination of uric acid in human urine and serum by capillary electrophoresis with chemiluminescence detection. *Anal. Biochem.*, *378*, 127-131.

Zheng, W., Wang, S., Chen, X., Hu, Z. (2004). Identification and determination of active anthraquinones in Chinese teas by micellar electrokinetic capillary chromatography. *Biomed. Chromatogr.*, *18*, 167-172.

Zhihong, L., Bo, G., Shuni, L., Mancheng, H., Shuping, X. (2004). Raman spectroscopic analysis of supersaturated aqueous solution of $MgO.B_2O_3$-32% $MgCl_2-H_2O$ during acidification and dilution. *Spectrochim. Acta A Mol. Biomol. Spectrosc.*, *60*, 3125-3128.

Zhu, H.Z., Cui, Y.M., Zheng, X.W., Han, H.R., Yang, M.M. (2007). Determination of trace trichlorfon by high performance liquid chromatography with UV detection based on its catalytic effect on sodium perborate oxidizing benzidine. *Anal. Chim. Acta*, *584*, 166-171.

Zinellu, A., Carru, C., Sotgia, S., Deiana, L. (2004). Optimization of ascorbic and uric acid separation in human plasma by free zone capillary electrophoresis ultraviolet detection. *Anal. Biochem.*, *330*, 298-305.

Zinellu, A., Pinna, A., Zinellu, E., Sotgia, S., Deiana, L., Carru, C. (2008). High-throughput capillary electrophoresis method for plasma cysteinylglycine measurement: evidences for a clinical application. *Amino Acids*, *34*, 69-74.

Chapter 3

PHARMACEUTICAL ASPECTS OF BORATES

Borate compounds such as boric acid, sodium borate and sodium perborate are mild antiseptics and are used in various pharmaceutical formulations. Boric acid is used as an antimicrobial preservative in eye drops, ointments and topical creams. Sodium borate has been used in mouth washes, otic preparations and eye drops. It is also used as an emulsifier in creams. Boric acid and sodium borate are used as buffering agents for eye drop preparations. Sodium perborate is used as a bleaching agent for dentifrices (Sweetman, 2009; Rowe *et al.*, 2009). The various characteristics of medicinally important borates and their clinical applications are described in the following sections.

3.1. CHARACTERISTICS OF MEDICINALLY IMPORTANT BORATES

The various characteristics of medicinally important borates, boric acid, sodium borate and sodium perborate, are described in British Pharmacopoeia (2009), European Pharmacopoeia (2005), United States Pharmacopeia (2007), Japanese Pharmacopoeia (2006), Martindale: The Complete Drug Reference (Sweetman, 2009), Handbook of Pharmaceutical Excipients (Rowe *et al.*, 2009) and Merck Index (O'Neil, 2001), and are presented as follows:

3.1.1. Boric Acid

Synonyms: Acidum boricum, boracic acid, boraic acid, boron trihydroxide, orthoboric acid, trihydroxyborene.
Empirical formula: H_3BO_3
Molecular structure:

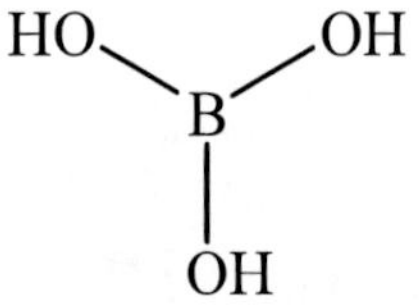

$$
\begin{array}{ccc}
HO & & OH \\
 & \diagdown \underset{|}{B} \diagup & \\
 & OH &
\end{array}
$$

Molecular weight: 61.8
CAS Registry number: 10043-35-3
Percent content: 99.0 to 100.5
Description: It occurs as colorless shiny plates or white or almost white crystals or crystalline powder.
Melting point: 170.9°C
Density: 1.435
pH (5% w/v aqueous solution): 3.5-4.1
pKa: 9.24
Solubility: One g is soluble in 18 mL cold, 4 mL of boiling water, in 18 mL cold, 6 mL of boiling ethanol and in 4 mL glycerol; solubility in water is increased by the addition of hydrochloric, citric or tartaric acids.
Pharmacopoeial Identification Tests
A. An acidic methanolic solution of boric acid, on ignition, burn with a green-bordered flame.
B. An aqueous solution of boric acid responds to the tests of borate.
Non-Pharmacopoeial Tests
C. Detection in biological material for cases of poisoning:
The blood or urine of the victim is acidified with concentrated hydrochloric acid and a drop of the sample is placed on the curcuma (turmeric) paper. After drying at room temperature, a red stain appears if boric acid is present. The detection limit is about 0.1 mg/mL (Yoshida *et al.*, 1989a)
D. Detection in tissues by histochemical staining:
Frozen 12-14 micron sections of the tissue, fixed in anhydrous alcohol, are stained in an acidified curcumin solution. After washing in acetic acid, a red stain indicates the presence of boric acid (Yoshida *et al.*, 1991).

Limit of sulfates: Maximum 450 ppm.
Limit of heavy metals: Maximum 15 ppm.
Limit of arsenic: Maximum 5 ppm.
Loss on drying: Not more than 0.5% (2 g, silica gel, 5 hours).
Stability and storage conditions: Boric acid volatilizes in steam. It is hygroscopic and should be stored in an air-tight container labeled "Not for Internal Use".
Toxicity: LD_{50} orally in rats: 5.14 g / kg.

3.1.2. Sodium Borate

Synonyms: Borax, boric and disodium salt, nitrii tetraboras, sodium biborate decahydrate, sodium pyroborate decahydrate, sodium tetraborate decahydrate.
Empirical formula: $Na_2B_4O_7, 10H_2O$
Molecular structure:

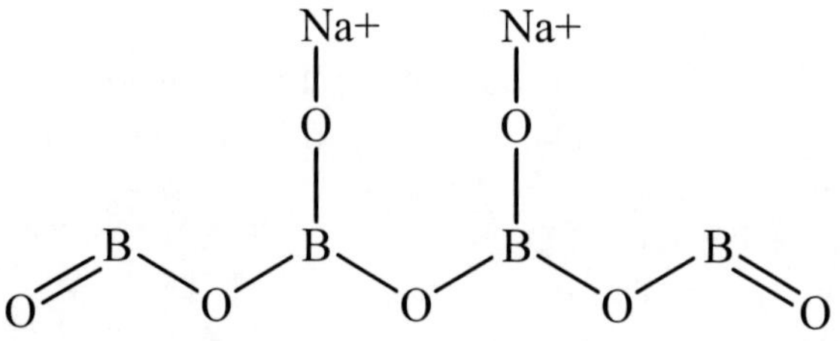

Molecular weight: 381.4
CAS Registry number: 1303-96-4
Percent content: 99.0 to 103.0
Description: It occurs as colorless or white crystals or a white crystalline powder.
Density: 1.73
Melting point: 75°C
pH (4% w/v aqueous solution): 9.0-9.6
Solubility: One g is soluble in 16 mL cold, 1 mL of boiling water, in 18 mL cold, 6 mL of boiling ethanol and in 1 mL of glycerol; insoluble in ethanol and diethyl ether.
Pharmacopoeial Identification Tests

A. An acidic methanolic solution of sodium borate, on ignition, burn with a green-bordered flame.
B. An aqueous solution of sodium borate, on the addition of phenolphthalein solution, produces a red color which disappears on the addition of glycerol.
C. An acidified aqueous solution of sodium borate, on the addition of few drops each of iodine solution and polyvinyl alcohol solution produces an intense blue color.

Limit of sulfates: Maximum 50 ppm.
Limit of heavy metals: Maximum 25 ppm.
Limit of arsenic: Maximum 5 ppm.
Limit of calcium: Maximum 100 ppm.
Stability and storage conditions: It should be stored in an air-tight container in a cool place.
Toxicity: LD_{50} orally in rats: 2.66 g / kg.

3.1.3. Sodium Perborate

Synonyms: hydrated sodium perborate, natrii perboras hydricus, natno perborate, natrium perboraatti, perborato sodico, perboritan sodny, sodium perborate de.
Empirical formula: $NaBH_2O_4, 3H_2O$
Molecular structure:

Molecular weight: 153.9
CAS Registry number: 7632-04-4
Percent content: 96.0 to 103.0
Description: It occurs as white or almost white crystalline powder.

Density: 1.73

Solubility: Sparingly soluble in water, dissolves in dilute mineral acid.

Pharmacopoeial Identification Tests

A. An acidic saturated solution of sodium perborate is mixed with potassium dichromate solution and shaken with diethyl ether. On standing the ether layer turns blue.

B. An acidic methanolic solution of sodium perborate, on ignition, burns with a green-bordered flame.

Limit of chlorides: Maximum 330 ppm.

Limit of sulfates: Maximum 1.2 percent.

Limit of heavy metals: Maximum 10 ppm.

Limit of iron: Maximum 20 ppm.

Stability and storage conditions: It should be stored in an air-tight container in a cool place.

3.2. CLINICAL USES

3.2.1. Antibacterial Agent

Boric acid and sodium borate are medicinally important compounds. Boric acid is weakly bacteriostatic and is used as a topical anti-infective in liquid dosage forms. It is non-irritating and its solutions are suitable for application to the cornea of the eye. Aqueous solutions of boric acid are employed as eyewash, mouthwash and for irrigation of the bladder. A 2.2% solution is isotonic with lacrimal fluid but such solutions will hemolyze red blood cells. It is also used as a dusting powder, on dilution with some inert material and can be absorbed through irritated skin. Sodium borate is bacteriostatic and has been employed as an ingredient of cold creams, eyewashes and mouthwashes. It is used as an alkalizing agent in denture adhesives. Sodium perborate is an oxidizing agent and is used as a local anti-infective. Various borate buffers are used to maintain the pH of alkaline pharmaceutical formulations (Soine and Wilson, 1967; O'Neil, 2001; Yelvigi, 2005; Reilly, 2006; Sweetman, 2009).

Combined treatment of purulent pyonecrotic lesions of lower extremities in diabetic patients has been performed by means of a permanent abacterial medium using 2% boric acid as an antiseptic (Shaposhnikov and Zorik, 2001). The minimum inhibitory concentration (MIC) of boric acid for bacteria is: Gram negative 800-12800 mg/L and Gram positive 1600-6400 mg/L (Grzybowska *et al.*, 2007).

Urine samples collected at home and in hospital have been preserved with boric acid for storage before processing (Porter and Brodie, 1969; Lum and Meers, 1989; Jewkes *et al.*, 1990; Meers and Chow, 1990; Lee and Critchley, 1998; Gillespie *et al.*, 1999; Thongboonkerd and Saetun, 2007; Thierauf, 2008; Kouri *et al.*, 2008; McDonald *et al.*, 2009; Wood, 2009; Thue *et al.*, 2010). Estrogen metabolites in breast cancer risk, excreted in urine, are preserved with boric acid before assay with an ELISA kit (Falk *et al.*, 2000). A preservation medium composed of nutrient broth, 1.8% boric acid, and 1% sodium chloride at pH 7.0 has been described to maintain the stability of *Escherichia coli* cultures for up to 10 days at room temperature (Brodsky *et al.*, 1978).

Phenylmercuric borate is used as an alternative antimicrobial preservative to phenylmercuric acetate or phenylmercuric nitrate in local anesthetic preparations and eye drops. It is more soluble in water than phenylmercuric nitrate and is less irritant than either phenylmercuric acetate or phenylmercuric nitrate (Abdelaziz and El-Nakeeb, 1988; Kodym *et al.*, 2003; British Pharmacopoeia, 2009).

3.2.2. Antifungal Agent

Boric acid is a fungistatic agent and has been used in the treatment of chronic vulvovaginal candidiasis (Van Slyke *et al.*, 1981; Jovanovic *et al.*, 1991; Prutting and Cerveny, 1998; Guaschino *et al.*, 2001; Vazquez *et al.*, 2004; Romano *et al.*, 2005; Ray *et al.*, 2007a,b; Das Neves *et al.*, 2008). It is effective against infections caused by various Candida species (*Candida albicans, Candida glabrata, Candida krusei, Candida parapsilosis*) (Sobel and Chaim, 1997; Otero *et al.*, 1999; Singh *et al.*, 2002; Sobel *et al.*, 2003; Van Kessel *et al.*, 2003; Romano *et al.*, 2005; Nyirjesy *et al.*, 2005; De Seta *et al.*, 2009), non-*Candida albicans* (Sood *et al.*, 2000; Otero *et al.*, 2002), Trichosporon species (Makela *et al.*, 2003), *Trichomonas vaginalis* (Aggarwal and Shier, 2008), and *Aspergillus niger* (Avino-Martinez *et al.*, 2008). The use of boric acid in suppressive antimicrobial therapy for recurrent bacterial vaginosis has been reported (Reichman *et al.*, 2009). In boric acid therapy the inhibition of oxidative metabolism appears to be a key factor in antifungal mechanism, but inhibition of virulence probably contributes to therapeutic efficacy in vivo (De Seta *et al.*, 2009). Boric acid and zinc oxide ointments have been used in the treatment of chronic eczema and psoriasis (Kubota *et al.*, 1983; Limaye and Weightman, 1997; Dallimore, 1998), trophic ulcers

(Izmailov and Izmailov, 1998), trabecular bone quality (Sheng *et al.*, 2001), dermatophytoses (Patel and Agrawal, 2002), foot epidermomycosis (Grosshans *et al.*, 1986), acute eczema (Bai *et al.*, 2007) and neural morphallaxis (Martinez *et al.*, 2008). As a fungistatic agent, boric acid is used in the form of a 2.5% aqueous-alcoholic solution (30:70, v/v), vaginal suppositories (600 mg), intravaginal capsules (600 mg) and topical creams.

3.2.3. Chemopreventive Agent for Human Cancer

Toxicity and carcinogenicity studies of boric acid in male and female mice have indicated that this is a noncarcinogenic chemical (Dieter, 1994). Boric acid and boronated compounds are used as boron carriers for boron neutron capture therapy in the treatment of glioblastoma, melanoma and other conditions. The suitability of these compounds in the treatment of cancer has been evaluated on the basis of pharmacokinetic studies characterizing their biodistribution, tumor uptake and tumor selectivity (Laster *et al.*, 1991; Gregoire *et al.*, 1993) and the effect of electroporation on cell killing (Ono *et al.*, 1998a,b; Kinashi *et al.*, 2002). The effectiveness of boron neutron capture in killing tumor cells depends on the number of ^{10}B atoms delivered to the tumor, the subcellular distribution of ^{10}B and the thermal neutron fluence at the side of the tumor. The presence of 600 ppm ^{10}B (boric acid) in the cell medium during irradiation with d(14) + Be neutrons enhances the DNA damage by 20% compared to that of the neutron irradiation alone (Poller *et al.*, 1996). Boric acid inhibits the proliferation of fibroblastoid murine L929-cells in a linear dose-dependent manner (Walmod *et al.*, 2004). Bor-tezomib, a dipeptide boronate proteasome inhibitor, shows activity in the treatment of multiple myeloma (Liu *et al.*, 2008). The use of L-boronophenyl alanine, sodium borocaptate and boric acid as boron carriers for boron neutron capture therapy in human hepatoma has been investigated in both hepG2 and clone 9 cells. Treatment with L-boronophenyl alanine had similar surviving fraction as those treated with boric acid after neutron irradiation (Chou *et al.*, 2009).

A calibration method is described for the determination of boron captured dose in boron neutron capture therapy using paired ion chambers. The calibration factor of the boronated ion chamber was determined as $1.83 \times 10^9 \pm$ 5.5% (± 1 sigma) n cm^{-2} per nC at standard temperature and pressure conditions (Wang *et al.*, 2007). Lithium-gadolinium-borate dispersed as microcrystals within the plastic scintillator BC-490 has been found as a promising material for accurate neutron dosimetry in mixed n/gamma fields (Lewis *et al.*, 2007).

Boric acid and 3-nitrophenylboronic acid have been found to inhibit human prostate cancer cell proliferation. Studies using DU-145 prostate cancer cells showed that boric acid induces a cell death-independent proliferative inhibition, with little effect on cell cycle stage distribution and mitochondrial function. It causes a dose-dependent reduction in cyclins A-E, as well as MAPK proteins, suggesting their contribution to proliferative inhibition. Boric acid is involved in the inhibition of the enzymatic activity of prostate-specific antigen which is a well-established marker of prostate cancer (Gallardo-Williams *et al.*, 2003, 2004; Barranco and Eckhert, 2004, 2006). It also inhibits NAD^+ and $NADP^+$ as well as the release of stored Ca^{2+} in growing DU-145 prostate cancer cells and thus impairs Ca^{2+} signaling (Barranco *et al.*, 2009). Boric acid causes a decrease of Ca^{2+} release from ryanodine receptor sensitive stores (Henderson *et al.*, 2009). Exposure to boric acid and calcium fructoborate results in the inhibition of the proliferation of breast cancer cells in a dose-dependent manner (Scorei *et al.*, 2008). Phenylboronic acid selectively inhibits human prostrate and breast cancer cells migration and decreases viability (Bradke *et al.*, 2008). Increased ground water concentrations of boric acid correlate with reduced risk of prostate cancer incidence and mortality. It also improves the anti-proliferative effectiveness of chemopreventive agents, selenomethionine and genistein, which enhances the ionization radiation effect on cell kill (Barranco *et al*, 2007).

3.2.4. Wound Treatment

Boric acid has been found to play an important role in wound healing. A 3% boric acid solution has been used in the treatment of deep wounds with loss of tissue. Dramatic improvement in wound healing was observed through its action on the extracellular matrix. Boron derivatives (triethanolamine borate, N-diethyl- phosphoramidate- propylboronic acid, 2,2-dimethylhexyl-1,3- propanediol- aminopropylboronate and 1,2- propanediol-aminopropylboronate) tested in a study mimicked the effects of boric acid (Blech *et al.*, 1990; Borrelly *et al.*, 1991; Benderdour *et al.*, 2000). A mixture consisting of aqueous boric acid and calcium hypochlorite solutions is widely used in the management of open wound healing (Salphale and Shenoi, 2003).

The presence of boron has been found to decrease the synthesis of proteoglycan, collagen and total proteins in pelvic cartilage of chick embryo but increases the release of these macromolecules (Benderdour *et al.*, 1997). The amount of phosphorylated proteins is enhanced in the presence of boron while endoprotease activity in cartilage is significantly increased. The *in vitro* effects of boric acid may explain its *in vivo* effects on wound healing (Benderdour *et al.*, 1997). Boric acid modulates extracellular matrix and tumor necrosis factor-alpha (TNF-alpha) synthesis in human fibroblasts. Total mRNA levels are higher after boric acid treatment and reach a maximum after 6 h. The effects of boric acid observed in wound repair may be due to TNF-alpha synthesis and secretion (Benderdour *et al.*, 1998). A lime solution containing boric acid as a dressing agent is better than sugar in the treatment of traumatic wounds in terms of contraction of size of wound, presence of discharge and formation of healthy granulation (Bajaj *et al.*, 2009). Rectovestibular fistula on simple resection or perineal dissection has been treated with 3% boric acid (Li *et al.*, 2010).

3.2.5. Eye Infections

Boric acid solution (11%) has been used in the treatment of ocular emergencies (Carvalho *et al.*, 2009). Acute corneal barrier changes induced by topically applied preservatives including boric acid (2%) have been evaluated using corneal transepithelial electric resistance in vivo (Kusano *et al.*, 2010).

Boric acid containing rewetting drops has been associated with a significant reduction in lens movement with a concurrent increase in lens centre thickness and reduction in postlens tear film thickness (Nicholas *et al.*, 2008). The chronic use of multipurpose contact lens care solutions containing boric acid by hydrogel contact lens wearers may lead to increase risk for associated corneal infection with *Pseudomonas aeruginosa* (Imayasu *et al.*, 2009a) and may increase adhesiveness of *A. castellanii* to corneal epithelial cells (Imayasu *et al.*, 2009b). The eye drops containing 0.0025% or lower concentrations of sodium perborate have been found to cause some degree of ocular tissue damage (Epstein *et al.*, 2009).

The corneal toxicity of two commercial anti-glaucoma ophthalmic solutions containing boric acid and benzalkonium chloride has shown that boric acid is less toxic to rat corneal epithelium than benzalkonium chloride (Nagai *et al.*, 2010). Borate-buffered hydrogel contact lens packaging solutions have been shown to adversely affect the viability and integrin expression of human corneal epithelial cells *in vitro* (Gorbet *et al.*, 2010). Contamination in the bottles of boric acid solution (40%) used in ophthalmology emergency has been detected due to inappropriate handling (Jose *et al.*, 2007).

3.2.6. Ear Infections

Boric acid has been used effectively as a topical antiseptic for treating asteatosis (Karakus *et al.*, 2003), otomycosis (Del Palacio *et al.*, 2002; Ozcan *et al.*, 2003), otits externa (Mendelsohn *et al.*, 2005) and chronic suppurative otitis media in children (Gao *et al.*, 1999; Macfadyen *et al.*, 2005; Minja *et al.*, 2006). A 4% aqueous solution of boric acid has been found to produce ototoxic effects when applied to the middle ear of guinea pigs (Ozturkcan *et al.*, 2009). Successful treatment of otitis externa associated with *Corynebacterium kroppenstedtii* in a lovebird (*Agapornis roseicollis*) with an acetic and boric acids solution has been reported (Martel *et al.*, 2009).

3.2.7. Dental Treatment

Hydrated sodium tetraborate (borax) is a component of glass-ionomer restorative cements in dentistry for its useful properties (Wilson and Kent, 1972; Bansal *et al.*, 1995). Boric acid is also a component of dental cements, commonly incorporated into glass (as an ingredient in the melt) and occasionally added to the powder component of the glass-ionomer cement (Prentice *et al.*, 2006). Boric acid as HBO_3^{2-} or BO_3^{3-}, may act as a weak polyalkenoate cross-linker in the acid-base glass-ionomer reaction involved in the setting process (Bochek *et al.*, 2002). Sodium perborate (30-35%) has been used as a bleaching agent for intracoronal bleaching of teeth (Teixeira *et al.*, 2003, 2004; Shinohara *et al.*, 2004, 2005; Timpawat *et al.*, 2005; Piemjai and Surakompontorn, 2006; Amaral *et al.*, 2008; Gokay *et al.*, 2008; Cavalli *et al.*, 2009; De Souza-Zaroni *et al.*, 2009), root filled discolored teeth (Lim *et al.*, 2004), artificially discolored teeth (Lee *et al.*, 2004; Baumler *et al.*, 2006), artificially stained primary teeth (Campos *et al.*, 2007; Arikan *et al.*, 2009), non-vital discolored teeth (Plotino *et al.*, 2008; Valera *et al.*, 2009), root filled teeth (Yui *et al.*, 2008) and blood stained teeth (Yui *et al.*, 2008). There is no significant effect of sodium perborate on the bond strength, hardness, and degree of infiltration between acrylic resin / resilient denture liners (Machado *et al.*, 2009; Pisani *et al.*, 2009). The use of chlorhexidine as a vehicle for sodium perborate enhances its antimicrobial activity (Oliveira *et al.*, 2008). A

3.8% sodium perborate solution could be used for the disinfection of acrylic dentures (Da Silva *et al.*, 2008). A 12 month color stability of enamel dentine, and animal-dentine samples after bleaching with a mixture of sodium perborate/distilled water has been reported. Color change of in vitro bleached tooth samples is not stable over time with regard to lightness. However, yellowness did not return to baseline within one year (Wiegand *et al.*, 2008).

3.3. Effect on Drug Stability

3.3.1. Catalytic Effect

Borate buffers have been used to maintain pH in the study of the stability of drug substances. Alterations in pH may affect the kinetics and mode of a chemical reaction (DeRitter, 1982; Connors *et al.*, 1986a; Sinko, 2006). Borate buffers act as a catalyst in the degradation of a number of drugs. In the alkaline borax-boric acid medium, a 0.5% atropine solution degraded to the extent of 44% in one month (Zvirblis *et al.*, 1956). Borate ions have a catalytic effect on the benzylpenicillinate ion in the degradation of benzylpenicillin (Hou and Poole, 1971). Carbenicillin undergoes general-base catalyzed hydrolysis in the presence of borate ions (Wood, 1986). Borate buffer increases the rate of degradation of cefotaxime in alkaline solution (Fabre *et al.*, 1984). The dependence of the observed rate constants for cephradine hydrolysis on total buffer concentration in borate buffer has been shown (Yamana and Tsuji, 1976). Enhanced oxidative degradation of hydrocortisone in borate and phosphate buffers appears to be due to trace metals since the catalytic effects can be greatly reduced by the addition of disodium edetate (Hansen and Bundgaard, 1979). The apparent activation energy for the degradation of hydrocortisone in 0.2 M borate buffer (pH 9.1) is 14.3 kcal mol^{-1} compared to that of 23.3 kcal mol^{-1} in 1.0 M HCl solution (Fleisher, 1986). The catalytic effect of borate buffer on the alkaline hydrolysis of indomethacin (pH 8.5) is about two fold compared to that of the phosphate buffer (McNamara, 1986). In aqueous methotrexate solutions, borate buffer may catalyze the hydrolysis of the drug in a concentration-dependent manner (Hansen *et al.*, 1983). The hydrolysis of oxytetracycline is subject to general-base catalysis by borate ions at pH 8.5-9.9 (Aminah, 1986). First-order rate constants for borate catalyzed hydrolysis of phenylbutazone at pH 9.1 have been determined. The rate of hydrolysis is slightly greater than that of the hydrolysis product, N-(2-carboxycaproyl)-hydrazobenzene (Schmid, 1970). The degradation of minocycline both under aerobic and anaerobic conditions is catalyzed by borate buffer (Pawelczyk and Matlak, 1982). The advantage of using borate or phosphate buffer in high-temperature sterilization of procaine solutions is much less because of the rapid change in hydroxyl ion concentration with temperature in these buffers (Higuchi and Busse, 1950). The activation energies for the alkaline hydrolysis of 5-fluorouracil in borate buffer are higher than those in the unbuffered solutions (Garrett et al., 1968). The methacholine chloride solutions prepared in borate buffer (pH 9.0) and stored at 27°C are degraded up to 60% after one week (Watson et al., 1998). Methanolic solutions of danazol in aqueous borate buffer undergo base-catalysis degradation by proton abstraction and follow pseudo-first order kinetics (Gadkariem et al., 2003). The degradation of Octastatin, a cyclic octapeptide, is influenced by borate buffer (Jang et al., 1997).

3.3.2. Stabilizing Effect

Borate buffer has been found to exert a stabilizing effect on chloramphenicol solutions whereas phosphate and citrate buffers catalyze the hydrolysis of the drug (Pandit, 1986). Boric acid is useful in ophthalmic formulations because of its physiological compatibility, its utility for adjustment of isotonicity and its stabilizing action on epinephrine (Connors, 1986b). The oxidation of α-methyldopa solutions is inhibited by the addition of borate ions (Sassetti and Fudenberg, 1971). Cocaine hydrochloride is incompatible with sodium borate, which liberates the free-base cocaine in alkaline solution (Connors et al., 1986a).

Under acidic conditions boric acid increases the thermostability of ribose and under basic conditions that of glucose (Scorei and Cimpoiasu, 2006). It is a useful preservative for the stabilization of ethyl glucuronide in urine samples that is employed as a specific and sensitive marker of ethanol consumption (Thierauf et al., 2008). Complexation with borates stabilizes ribose and the ribose-borate complex is more stable than those of arabinose, lyxose and xylose (Sponer et al., 2008).

Borate buffer is known to complex with the ribityl side chain of riboflavin (vitamin B_2) in a reversible 1:1 association and the negatively charged complex is more resistant to hydroxyl ion attack on the isoalloxazine ring (Wadke and Guttman, 1964). The boric acid component of the buffer is involved in the formation of the riboflavin-borate complex (Dudutz and Ristea, 1978; Pinto and Rivlin, 1987; Ruth et al., 1991; Rivlin, 2007), which could influence the rate of its photolysis reactions (Ahmad *et al.*, 2004; Ahmad and Vaid, 2006). Recently a detailed study of the photolysis of riboflavin at 8.0-10.5 in the presence of borate buffer has been conducted and the rate constants for the reaction have been determined (Ahmad *et al.*, 2008). The values of the rate constants have been found to decrease with an increase in buffer concentration suggesting that the borate species inhibits the rate of photolysis of riboflavin as a result of the formation of riboflavin-borate complex. The rate constants for the photolysis of riboflavin are about two times slower than those obtained in the absence of borate buffer indicating a significant buffer effect on the reaction.

3.4. DISSOLUTION KINETICS

Robson *et al.* (2000) studied the influence of buffer composition on the release of cefuroxime axetil from stearic acid micropheres with particular emphasis on establishing the relationship between buffer composition and drug release. Drug dissolution and release from microspheres at pH 7.0 indicated marked differences in release profile with an approximate rank order of Sorensens modified phosphate buffer > citrate phosphate buffer > approximately boric acid buffer > mixed phosphate buffer.

The dissolution of probertite ($NaCaB_5O_9.5H_2O$) in boric acid solution has been investigated as a function of temperature. The increase in the boric acid concentration leads to an increase in the dissolution of probertite. However, the boric acid concentration above 5% at 60 and 80°C does not significantly affect the dissolution of probertite. The dissolution kinetics of probertite follows pseudo first-order reaction (Mergen and Demirhan, 2009).

3.5. PHARMACOKINETICS

It has been demonstrated that water-emulsifying and hydrophobic ointments containing boric acid liberate only minute amounts (1–6%) within 24 h compared with the nearly total liberation from a jelly. The pharmacokinetics rules out the risk of cumulative poisoning with topical preparations containing low amounts (up to 3%) of boric acid (Schon *et al.*, 1984). A pharmacokinetic study has been conducted by giving to eight young adult males a single dose of boric acid (562–611mg) by I.V. infusion. The plasma concentration curves best fitted a three-compartment open model. The 120 h urinary excretion was $98.7 \pm 9.1\%$ of dose, C_{tot} 54.6 ± 8.0 mL/min/1.73 m^2, $t_{1/2}$ beta 21.0 ± 4.9 h and distribution volumes V_1, V_2 and V_3: 0.251 ± 0.099, 0.456 ± 0.067 and 0.340 ± 0.128 L/kg, respectively (Jansen *et al.*, 1984). Suicidal ingestion of about 21 g of boric acid by a young female showed the concentration of boric acid in serum and urine as 465 µg/mL and 3.40 mg/mL, respectively. The half-life of boric acid in serum (13.46 h) was decreased to 3.76 h and the total body clearance (0.99 L/h) increased to 3.53 L/h on hemodialysis. The additional removal of boric acid by the method was about 5 g. Thus hemodialysis has been found to be very useful in the treatment of acute boric acid poisoning (Teshima *et al.*, 1992; Naderi and Palmer, 2006).

The pharmacokinetics of sodium tetraborate in rats by administering 1 mL oral dose to several groups of rats (n=20) at several dose levels ranging from 0-0.4 mg/100 g body weight as boron has been studied. After 24 h the average urinary excretion rate for the element was $99.6 \pm 7.9\%$. The data have been interpreted according to a one-compartment open model. The various parameters estimated are: absorption half-life, $t_{1/2}$ α 0.608 ± 0.432 h; elimination half-life, $t_{1/2}$ 4.64 ± 1.19 h; volume of distribution, V_d 142.0 ± 30.2 mL/100 g body weight; total clearance, C_{tot} 0.359 ± 0.0285 mL/min/100 g body weight. The maximum boron concentration (C_{max}) in serum after administration was 2.13 ± 0.270 mg/L, and the time needed to reach the maximum concentration (T_{max}) was 1.76 ± 0.887 h. The pharmacokinetic model is proposed as a useful tool to deal with the problem of environmental or industrial exposure to boron or in the case of accidental acute intoxication (Usuda *et al.*, 1998). The half-life of boric acid in humans is on the order of one day (Moseman, 1994). A HPLC method for the determination of linezolid in pulmonary tissue has been reported. The tissue samples are prepared in Tris-boric acid-EDTA buffer and the method can be used in pharmacokinetic studies (Guerrero *et al.*, 2010). A comparative review of the pharmacokinetics of boric acid in humans and rodents shows remarkable similarity in their behavior (Murray, 1998).

The [(18) F]–labeled aryltrifluoroborate has been found to clear *in vivo* quite rapidly to the bladder without leaching of free [(18) F]–fluoride to the bone. When the labeled biotinylated boronic ester is preincubated with avidin, the pharmacokinetic clearance of the resulting complex is altered. The study suggest that boronic esters are potentially useful as readily labeled precursors to [(18)F]–positron emission tomography (PET) reagents (Ting *et al.*, 2008).

Boric acid ingestion has been associated with greatly increased urinary excretion of riboflavin in patients, both children and adults. Most of the riboflavin is excreted within the first 24 h after ingestion of boric acid (Pinto *et al.*, 1978). Boric acid complexes with the polyhydroxyl ribitol side chain of riboflavin and causes enhanced excretion of the vitamin to promote riboflavinuria in animals and man (Pinto and Rivlin, 1987).

3.6. YEAST GROWTH

Boron is required for the growth of vascular plants and stimulates yeast (*Saccharomyces cerevisiae*) growth. It could be used as a model for the evaluation of intracellular boron trafficking (Bennet *et al.*, 1999). The role of three membrane proteins (BOR1, DUR3 and FPS1) in boron transport in yeast has been examined. The boron concentration in yeast cells lacking BOR1 is elevated upon exposure to 90 mM boric acid, whereas cells lacking DUR3 or FPS1 showed lower boron concentrations. Thus, three proteins appear to be involved in tolerance of boric acid and the maintenance of the protoplasmic boron concentration (Nozawa *et al.*, 2006a). *Arabidopsis thaliana* cDNA genes have been isolated that confer boric acid tolerance on wild-type yeast (Nozawa *et al.*, 2006b).

The functional properties of the *Saccharomyces cerevisiae* bicarbonate transporter homolog Bor1p (YNL275wp) have been characterized by measuring boron (as boric acid) and sodium and chloride ion fluxes. Neither sodium nor chloride appears to be a transported substrate for Bor1p. Uphill efflux of boron mediated by Bor1p was demonstrated directly by loading cells with boron and resuspending in a low-boron medium. Cells with intact BOR1 transport boron outward until the intracellular concentration is sevenfold lower than that in the medium (Jennings *et al.*, 2007).

The Atr1 Delta mutants are highly sensitive to boron treatment, whereas cells overexpressing ATR1 are boron resistant. The Atr1 functions as a boron efflux pump and is required for boron tolerance in yeast (Kaya *et al.*, 2009).

REFERENCES

Abdelaziz, A.A., El-Nakeeb, M.A. (1988). Sporicidal activity of local anesthetics and their binary combinations with preservatives. *J. Clin. Pharm. Ther.*, *13*, 249-256.

Aggarwal, A., Shier, R.M. (2008). Recalcitrant *Trichomonas vaginalis* infections successfully treated with vaginal acidification. *J. Obstet. Gynaecol. Can.*, *30*, 55-58.

Ahmad, I., Vaid, F.H.M. (2006). Photochemistry of flavins in aqueous and organic solvents, In: E. Silva, A.M. Edwards, Eds., *Flavins: Photochemistry and Photobiology*, Cambridge: Royal Society of Chemistry, pp. 13-40.

Ahmad, I., Fasihullah, Q., Noor, A., Ansari, I.A., Ali, Q.N.M. (2004). Photolysis of riboflavin in aqueous solution: a kinetic study. *Int. J. Pharm.*, *280*, 199-208.

Ahmad, I., Ahmed, S., Sheraz, M.A., Vaid, F.H.M. (2008). Effect of borate buffer on the photolysis of riboflavin in aqueous solution. *J. Photochem. Photobiol. B: Biol.*, *93*, 82-87.

Amaral, C., Jorge, A., Veloso, K., Erhardt, M., Arias, V., Rodrigues, J.A. (2008). The effect of in-office in combination with intracoronal bleaching on enamel and dentin bond strength and dentin morphology. *J. Contemp. Dent. Pract.*, *9*, 17-24.

Aminah, Rd. (1986). Oxytetracycline monograph, In: K.A. Connors, G.L. Amidon, V.J. Stella, Eds., *Chemical Stability of Pharmaceuticals: A Handbook for Pharmacists*, 2nd ed., New York, NY: John Wiley & Sons, and references therein, pp. 463-468.

Arikan, V., Sari, S., Sonmez, H. (2009). Bleaching a devital primary tooth using sodium perborate with walking bleach technique: a case report. *Oral Surg. Oral Med. Oral Pathol. Oral Radiol. Endod.*, *107*, e80-84.

Avino-Martinez, J.A., Espana-Gregori, E., Peris-Martinez, C.P., Blanes, M. (2008). Successful boric acid treatment of Aspergillus niger infection in an exenterated orbit. *Ophthal. Plast. Reconstr. Surg.*, *24*, 79-81.

Bai, Y.P., Yang, D.Q., Wang, Y.M. (2007). Clinical study on treatment of acute eczema by Shuangfujin. *Zhongguo Zhong Xi Yi Jie He Za Zhi.*, *27*, 72-75.

Bajaj, G., Karn, N.K., Shrestha, B.P., Kumar, P., Singh, M.P. (2009). A randomized controlled trial comparing eusol and sugar as dressing agents in the treatment of traumatic wounds. *Trop. Doct.*, *39*, 1-3.

Bansal, R.K., Tewari, U.S., Singh, P., Murthy, D.V.S. (1995). Modified polyalkenoate (glass-ionomer) cement-a study. *J. Oral Rehabil.*, *22*, 533-537.

Barranco, W.T., Eckhert, C.D. (2004). Boric acid inhibits human prostate cancer cell proliferation. *Cancer Lett.*, *216*, 21-29.

Barranco, W.T., Eckhert, C.D. (2006). Cellular changes in boric acid-treated DU-145 prostate cancer cells. *Br. J. Cancer*, *94*, 884-890.

Barranco, W.T., Hudak, P.F., Eckhert, C.D. (2007). Evaluation of ecological and in vitro effects of boron on prostate cancer risk (United States). *Cancer Causes Control*, *18*, 71-77.

Barranco, W.T., Kim, D.H., Stella, S.L. Jr., Eckhert, C.D. (2009). Boric acid inhibits stored Ca^{2+} release in DU-145 prostate cancer cells. *Cell Biol. Toxicol.*, *25*, 309-320.

Bäumler, M.A., Schug, J., Schmidlin, P., Imfeld, T. (2006). In vitro tests of internal tooth whitening agents on colored solutions do not replace tests on teeth. *Schweiz. Monatsschr. Zahnmed.*, *116*, 1000-1005.

Benderdour, M., Hess, K., Dzondo-Gadet, M., Nabet, P., Belleville, F., Dousset, B. (1998). Boron modulates extracellular matrix and TNF alpha synthesis in human fibroblasts. *Biochem. Biophys. Res. Commun.*, *246*, 746-751.

Benderdour, M., Hess, K., Gadet, M.D., Dousset, B., Nabet, P., Belleville, F. (1997). Effect of boric acid solution on cartilage metabolism. *Biochem. Biophys. Res. Commun.*, *234*, 263-268.

Benderdour, M., Van Bui, T., Hess, K., Dicko, A., Belleville, F., Dousset, B. (2000). Effects of boron derivatives on extracellular matrix formation. *J. Trace Elem. Med. Biol.*, *14*, 168-173.

Bennett, A., Rowe, R.I., Soch, N., Eckhert, C.D. (1999). Boron stimulates yeast (*Saccharomyces cerevisiae*) growth. *J. Nutr.*, *129*, 2236-2238.

Blech, M.F., Martin, C., Borrelly, J., Hartemann, P. (1990). Treatment of deep wounds with loss of tissue. Value of a 3 percent boric acid solution. *Presse Med. 19*, 1050-1052.

Bochek, A.M., Yusupova, L.D., Zabivalova, N.M., Petropavlovskii, G.A. (2002). Rheological properties of aqueous H-carboxymethyl cellulose solutions with various additives. *Russ. J. Appl. Chem.*, *75*, 645-648.

Borrelly, J., Blech, M.F., Grosdidier, G., Martin-Thomas, C., Hartemann, P. (1991). Contribution of a 3% solution of boric acid in the treatment of deep wounds with loss of substance. *Ann. Chir. Plast. Esthet.*, *36*, 65-69.

Bradke, T.M., Hall, C., Carper, S.W., Plopper, G.E. (2008). Phenylboronic acid selectively inhibits human prostate and breast cancer cell migration and decreases viability. *Cell Adh. Migr.*, *2*, 153-160.

British Pharmacopoeia (2009). London: Her Majesty's Stationary Office, Electronic Version.

Brodsky, M.H., Ciebin, B.W., Schiemann, D.A. (1978). Simple bacterial preservation medium and its application to proficiency testing in water bacteriology. *Appl. Environ. Microbiol.*, *35*, 487-491.

Campos, S.F., César, I.C., Munin, E., Liporoni, P.C., do Rego, M.A. (2007). Analysis of photoreflectance and microhardness of the enamel in primary teeth submitted to different bleaching agents. *J. Clin. Pediatr. Dent.*, *32*, 9-12.

Carvalho, R.S., Kara-José, N., Temporini, E.R., Kara-Junior, N., Noma-Campos, R. (2009). Self-medication: initial treatments used by patients seen in an ophthalmologic emergency room. *Clinics (Sao Paulo)*, *64*, 735-741.

Cavalli, V., Shinohara, M.S., Ambrose, W., Malafaia, F.M., Pereira, P.N., Giannini, M. (2009). Influence of intracoronal bleaching agents on the ultimate strength and ultrastructure morphology of dentine. *Int. Endod. J.*, *42*, 568-575.

Chou, F.I., Chung, H.P., Liu, H.M., Chi, C.W., Lui, W.Y. (2009). Suitability of boron carriers for BNCT: accumulation of boron in malignant and normal liver cells after treatment with BPA, BSH and BA. *Appl. Radiat. Isot.*, *67 (7-8 Suppl)*, S105-108.

Connors, K.A. (1986b). Epinephrine monograph, In: K.A. Connors, G.L. Amidon, V.J. Stella, Eds., *Chemical Stability of Pharmaceuticals: A Handbook for Pharmacists*, 2nd ed., New York, NY: John Wiley & Sons, and references therein, pp. 438-447.

Connors, K.A., Amidon, G.L., Stella, V.J., Eds. (1986a). *Chemical Stability of Pharmaceuticals: A Handbook for Pharmacists*, 2nd ed., New York, NY: John Wiley & Sons, pp. 43, 90, 376.

da Silva, F.C., Kimpara, E.T., Mancini, M.N., Balducci, I., Jorge, A.O., Koga-Ito, C.Y. (2008). Effectiveness of six different disinfectants on removing five microbial species and effects on the topographic characteristics of acrylic resin. *J. Prosthodont.*, *17*, 627-633.

Dallimore, K.J. (1998). Effect of an ointment containing boric acid, zinc oxide, starch and petrolatum on psoriasis. *Australas. J. Dermatol.*, *39*, 283.

Das Neves, J., Pinto, E., Teixeira, B., Dias, G., Rocha, P., Cunha, T., Santos, B., Amaral, M.H., Bahia, M.F. (2008). Local treatment of vulvovaginal candidosis: general and practical considerations. *Drugs, 68*, 1787-1802.

De Seta, F., Schmidt, M., Vu, B., Essmann, M., Larsen, B. (2009). Antifungal mechanisms supporting boric acid therapy of *Candida vaginitis. J. Antimicrob. Chemother., 63*, 325-336.

de Souza-Zaroni, W.C., Lopes, E.B., Ciccone-Nogueira, J.C., Silva, R.C. (2009). Clinical comparison between the bleaching efficacy of 37% peroxide carbamide gel mixed with sodium perborate with established intracoronal bleaching agent. *Oral Surg. Oral Med. Oral Pathol. Oral Radiol. Endod., 107*, e43-47.

Del Palacio, A., Cuetara, M.S., Lopez-Suso, M.J., Amor, E., Garau, M. (2002). Randomized prospective comparative study: short-term treatment with ciclopiroxolamine (cream and solution) versus boric acid in the treatment of otomycosis. *Mycoses, 45*, 317-328.

DeRitter, E. (1982). Vitamins in Pharmaceutical formulations. *J. Pharm. Sci., 71*, 1073-1096.

Dieter, M.P. (1994). Toxicity and carcinogenicity studies of boric acid in male and female B6C3F1 mice. *Environ. Health Perspect., 102*, 93-97.

Dudutz, G., Ristea, I. (1978). Physicochemical study of the interaction of boric acid with pantothenic acid and riboflavin. *Rev. Med. (Romania), 24*, 188-192.

Epstein, S.P., Ahdoot, M., Marcus, E., Asbell, P.A. (2009). Comparative toxicity of preservatives on immortalized corneal and conjunctival epithelial cells. *J. Ocul. Pharmacol. Ther., 25*, 113-119.

European Pharmacopoeia (2005). European Directorate for the Quality of Medicines, 5th ed., Council of Europe, Strasbourg, France.

Fabre, H., Eddine, N.H., Berge, G. (1984). Degradation kinetics in aqueous solution of cefotaxime sodium, a third-generation cephalosporin. *J. Pharm. Sci., 73*, 611-618.

Falk, R.T., Rossi, S.C., Fears, T.R., Sepkovic, D.W., Migella, A., Adlercreutz, H., Donaldson, J., Bradlow, H.L., Ziegler, R.G. (2000). A new ELISA kit for measuring urinary 2-hydroxyestrone, 16 alpha-hydroxyestrone, and their ratio; reproducibility, validity, and assay performance after freeze-thaw cycling and preservation by boric acid. *Cancer Epidemiol. Biomarkers Prev., 9*, 81-87.

Fleisher, D. (1986). Hydrocortisone monograph, In: K.A. Connors, G.L. Amidon, V.J. Stella, Eds., *Chemical Stability of Pharmaceuticals: A*

Handbook for Pharmacists, 2nd ed., New York, NY: John Wiley & Sons, and references therein, pp. 483-490.

Gadkariem, E.A., El-Obeid, H.A., Abounassif, M.A., Ahmad, S.M., Ibrahim, K.E. (2003). Effect of alkali and simulated gastric and intestinal fluids on danazol stability. *J. Pharm. Biomed. Anal., 26*, 743-751.

Gallardo-Williams, M.T., Maronpot, R.R., Wine, R.N., Brunssen, S.H., Chapin, R.E. (2003). Inhibition of the enzymatic activity of prostate-specific antigen by boric acid and 3-nitrophenyl boronic acid. *Prostate, 54*, 44-49.

Gallardo-Williams, M.T., Chapin, R.E., King, P.E., Moser, G.J., Goldsworthy, T.L., Morrison, J.P., Maronpot, R.R. (2004). Boron supplementation inhibits the growth and local expression of IGF-1 in human prostate adenocarcinoma (LNCaP) tumors in nude mice. *Toxicol. Pathol., 32*, 73-78.

Gao, Q., Liu, Z., Cui, Y. (1999). Clinical and bacteriology observation on intraoperation usage of tarivid otic solution in treatment of chronic otitis media with cholesteatoma. *Lin Chuang Er Bi Yan Hon Ke Za Zhi, 13*, 219-220.

Garrett, E.R., Nestler, H.J., Somodi, A. (1968). Kinetics and mechanisms of hydrolysis of 5-halouracils. *J. Org. Chem., 33*, 3460-3468.

Gillespie, T., Fewster, J., Masterton, R.G. (1999). The effect of specimen processing delay on borate urine preservation. *J.Clin. Pathol., 52*, 95-98.

Gökay, O., Ziraman, F., Cali Asal, A., Saka, O.M. (2008). Radicular peroxide penetration from carbamide peroxide gels during intracoronal bleaching. *Int. Endod. J., 41*, 556-560.

Gorbet, M.B., Tanti, N.C., Jones, L., Sheardown, H. (2010). Corneal epithelial cell biocompatibility to silicone hydrogel and conventional hydrogel contact lens packaging solutions. *Mol. Vis., 16*, 272-282.

Gregoire, V., Begg, A.C., Huiskamp, R., Veiryk, R., Bartelink, H. (1993). Selectivity of boron carriers for boron neutron capture therapy: pharmacological studies with borocapture sodium, L-boronophenylalanine and boric acid in murine tumors. *Radiother. Oncol., 27*, 46-54.

Grosshans, E., Schwaab, E., Samsoen, M., Grange, D., Koenig, H., Kremer, M. (1986). Clinical aspects, epidemiology and economic impact of foot epidermomycosis in an industrial milieu. *Ann. Dermatol. Venereol., 1113*, 521-533.

Grzybowska, W., Młynarczyk, G., Młynarczyk, A., Bocian, E., Luczak, M., Tyski, S. (2007). Estimation of activity of pharmakopeal disinfectants and antiseptics against Gram-negative and Gram-positive bacteria isolated from clinical specimens, drugs and environment. *Med. Dosw. Mikrobiol.,* *59*, 65-73.

Guaschino, S., Deseta, F., Sartore, A., Ricci, G., De Santo, D., Piccoli, M., Alberico, S. (2001). Efficacy of maintenance therapy with topical boric acid in comparison with oral itraconazole in the treatment of recurrent vulvovaginal candidiasis. *Am. J. Obstet. Gynecol., 184*, 598-602.

Guerrero, L., Martínez-Olondris, P., Rigol, M., Esperatti, M., Esquinas, C., Luque, N., Piner, R., Torres, A., Soy, D. (2010). Development and validation of a high performance liquid chromatography method to determine linezolid concentrations in pig pulmonary tissue. *Clin. Chem. Lab. Med., 48*, 391-398.

Hansen, J., Bundgaard, H. (1979). Studies on the stabilities of corticosteroids I. Kinetics of degradation of hydrocortisone in aqueous solution. *Arch. Pharm. Chem., Sci. Ed., 7*, 135-146.

Hansen, J., Kreilgard, B., Nielson, O., Veje, J. (1983). Kinetics of degradation of methotrexate in aqueous solution. *Int. J. Pharm., 16*, 141-152.

Henderson, K., Stella, S.L., Kobylewski, S., Eckhert, C.D. (2009). Receptor activated Ca^{2+} release is inhibited by boric acid in prostate cancer cells. *PLoS One, 4*, e6009.

Higuchi, T., Busse, L.W. (1950). Heat sterilization of thermally labile solutions. *J. Am. Pharm. Assoc. Sci. Ed., 39*, 411-412.

Hou, J.P., Poole, J.W. (1971). ß-lactam antibiotics: Their physicochemical properties and biological activities in relation to structure. *J. Pharm. Sci., 60*, 503-532.

Imayasu, M., Shimizu, H., Shimada, S., Suzuki, T., Cavanagh, H.D. (2009a). Effects of multipurpose contact-lens care solutions on adhesion of *Pseudomonas aeruginosa* to corneal epithelial cells. *Eye Contact Lens, 35*, 98-104.

Imayasu, M., Uno, T., Ohashi, Y., Cavanagh, H.D. (2009b). Effects of multipurpose contact lens care solutions on the adhesiveness of Acanthamoeba to corneal epithelial cells. *Eye Contact Lens, 35*, 246-250.

Izmailov, G.A., Izmailov, S.G. (1998). Treatment of trophic ulcers of lower extremities by use of boric acid-hydrocortisone mixture. *Khirurgiia (Mosk.), 1*, 46-47.

Jang, S.W., Woo, B.H., Lee, J.T., Moon, S.C., Lee, K.C., DeLuca, P.P. (1997). Stability of Octastatin, a somatostatin analog cyclic octapeptide, in aqueous solution. *Pharm. Dev. Technol., 2*, 409-414.

Jansen, J.A., Anderson, J., Schon, J.S. (1984). Boric acid single dose pharmacokinetics after intravenous administration to man. *Arch. Toxicol.*, *55*, 64-67.

Japanese Pharmacopeia (2006). Pharmacopoeia of Japan, 15th ed., Yakuji Nippo Ltd., Tokyo, Japan, Electronic Version.

Jennings, M.L., Howren, T.R., Cui, J., Winters, M., Hannigan, R. (2007). Transport and regulatory characteristics of the yeast bicarbonate transporter homolog Bor1p. *Am. J. Physiol. Cell Physiol.*, *293*, C468-476.

Jewkes, F.E., McMaster, D.J., Napier, W.A., Houston, I.B., Postlethwaite, R.J. (1990). Home collection of urine specimens-boric acid bottles or dipslides? *Arch. Dis. Child*, *65*, 286-289.

José, A.C., Castelo Branco, B., Ohkawara, L.E., Yu, M.C., Lima, A.L. (2007). Use conditions of boric acid solution in the eye: handling and occurrence of contamination. *Arq. Bras. Oftalmol.*, *70*, 201-207.

Jovanovic, R., Congema, E., Nguyen, H.T. (1991). Antifungal agents vs. boric acid for treating chronic mycotic vulvovaginitis. *J. Reprod. Med.*, *36*, 593-597.

Karakus, M.E., Arda, H.N., Ikinciogullari, A., Gedikli, Y., Coskun, S., Balaban, N., Akdogan, O. (2003). Microbiology of the external auditory canal in patients with asteatosis and itching. *Kulak. Burun. Bogaz. Ihtis. Derg.*, *11*, 33-38.

Kaya, A., Karakaya, H.C., Fomenko, D.E., Gladyshev, V.N., Koc, A. (2009). Identification of a novel system for boron transport: Atr1 is a main boron exporter in yeast. *Mol. Cell Biol.*, *29*, 3665-3674.

Kinashi, Y., Masunaga, S., Ono, K. (2002). Mutagenic effect of borocaptate sodium and boronophenylalanine in neutron capture therapy. *Int. J. Radiat. Oncol. Biol. Phys.*, *54*, 562-567.

Kodym, A., Marcinkowski, A., Kukuła, H. (2003). Technology of eye drops containing aloe (*Aloe arborescens* Mill-Liliaceae) and eye drops containing both aloe and neomycin sulphate. *Acta Pol. Pharm.*, *60*, 31-39.

Kouri, T., Malminiemi, O., Penders, J., Pelkonen, V., Vuotari, L., Delanghe, J. (2008). Limits of preservation of samples for urine strip tests and particle counting. *Clin. Chem. Lab. Med.*, *46*, 703-713.

Kubota, K., Kumakiri, M., Miura, Y., Hine, K., Kori, N., Saito, H., Miyazaki, K., Arita, T. (1983). Clinical studies on zinc oxide ointment replacing boric acid and zinc oxide ointment (JP8). *Hokkaido Igaku Zasshi*, *58*, 400-405.

Kusano, M., Uematsu, M., Kumagami, T., Sasaki, H., Kitaoka, T. (2010). Evaluation of acute corneal barrier change induced by topically applied preservatives using corneal transepithelial electric resistance in vivo. *Cornea*, *29*, 80-85.

Laster, B.H., Kahl, S.B., Popenoe, E.A., Pate, D.W., Fairchild, R.G. (1991). Biological efficacy of boronated low-density lipoprotein for boron neutron capture therapy as measured in cell culture. *Cancer Res.*, *51*, 4588-4593.

Lee, Z.S., Critchley, J.A. (1998). Simultaneous measurement of catecholamines and kallikrein in urine using boric acid preservative. *Clin. Chim. Acta*, *276*, 89-102.

Lee, G.P., Lee, M.Y., Lum, S.O., Poh, R.S., Lim, K.C. (2004). Extraradicular diffusion of hydrogen peroxide and pH changes associated with intracoronal bleaching of discoloured teeth using different bleaching agents. *Int. Endod. J.*, *37*, 500-506.

Lewis, D.V., Spyrou, N.M., Williams, A.M., Beeley, P.A. (2007). Lithium-gadolinium-borate as a neutron dosemeter. *Radiat. Prot. Dosimetry*, *126*, 390-393.

Li, L., Zhang, T.C., Zhou, C.B., Pang, W.B., Chen, Y.J., Zhang, J.Z. (2010). Rectovestibular fistula with normal anus: a simple resection or an extensive perineal dissection? *J. Pediatr. Surg.*, *45*, 519-524.

Lim, M.Y., Lum, S.O., Poh, R.S., Lee, G.P., Lim, K.C. (2004). An in vitro comparison of the bleaching efficacy of 35% carbamide peroxide with established intracoronal bleaching agents. *Int. Endod. J.*, *37*, 483-488.

Limaye, S., Weightman, W. (1997). Effect of an ointment containing boric acid, zinc oxide, starch and petrolatum on psoriasis. *Australas J. Dermatol.*, *38*, 185-186.

Liu, F.T., Agrawal, S.G., Movasaghi, Z., Wyatt, P.B., Rehman, I.U., Gribben, J.G., Newland, A.C., Jia, L. (2008). Dietary flavonoids inhibit the anticancer effects of the proteasome inhibitor bortezomib. *Blood*, *112*, 3835-3846.

Lum, K.T., Meers, P.D. (1989). Boric acid converts urine into an effective bacteriostatic transport medium. *J. Infect.*, *18*, 51-58.

Macfadyen, C., Gamble, C., Garner, P., Macharia, I., Mackenzie, I., Mugwe, P., Oburra, H., Otwombe, K., Taylor, S., Williamson, P. (2005). Topical quinolone vs. antiseptic for treating chronic suppurative otitis media: a randomized controlled trial. *Trip. Med. Int. Health*, *10*, 190-197.

Machado, A.L., Breeding, L.C., Vergani, C.E., da Cruz Perez, L.E. (2009). Hardness and surface roughness of reline and denture base acrylic resins after repeated disinfection procedures. *J. Prosthet. Dent.*, *102*, 115-122.

Makela, P., Leaman, D., Sobel, J.D. (2003). Vulvovaginal trichosporonosis. *Infect. Dis. Obstet. Gynecol.*, *11*, 131-133.

Martel, A., Haesebrouck, F., Hellebuyck, T., Pasmans, F. (2009). Treatment of otitis externa associated with *Corynebacterium kroppenstedtii* in a peachfaced lovebird (*Agapornis roseicollis*) with an acetic and boric acid commercial solution. *J. Avian Med. Surg., 23*, 141-144.

Martinez, V.G., Manson, J.M., Zoran, M.J. (2008). Effects of nerve injury and segmental regeneration on the cellular correlates of neural morphallaxis. *J. Exp. Zoolog. B. Mol. Dev. Evol., 310*, 520-533.

McDonald, T.J., Knight, B.A., Shields, B.M., Bowman, P., Salzmann, M.B., Hattersley, A.T. (2009). Stability and reproducibility of a single-sample urinary C-peptide/creatinine ratio and its correlation with 24-h urinary C-peptide. *Clin. Chem., 55*, 2035-2039.

McNamara, D. (1986). Indomethacin monograph, In: K.A. Connors, G.L. Amidon, V.J. Stella, Eds., *Chemical Stability of Pharmaceuticals: A Handbook for Pharmacists*, 2nd ed., New York, NY: John Wiley & Sons, and references therein, pp. 509-516.

Meers, P.D., Chow, C.K. (1990). Bacteriostatic and bactericidal actions of boric acid against bacteria and fungi commonly found in urine. *J. Clin. Pathol., 43*, 484-487.

Mendelsohn, C.L., Griffin, C.E., Rosenkrantz, W.S., Brown, L.D., Boord, M.J. (2005). Efficacy of boric-complexed zinc and acetic-complexed zinc otic preparations for canine yeast otitis externa. *J. Am. Anim. Hosp. Assoc., 41*, 12-21.

Mergen, A., Demirhan, M.H. (2009). Dissolution kinetics of probertite in boric acid solution. *Int. J. Miner. Process., 90*, 16-20.

Minja, B.M., Moshi, N.H., Ingvarsson, L., Bastos, I., Grenner, J. (2006). Chronic suppurative otitis media in Tanzanian school children and its effects on hearing. *East Afr. Med. J., 83*, 322-325.

Moseman, R.F. (1994). Chemical disposition of boron in animals and humans. *Environ. Health Perspect., 102*, 113-117.

Murray, F.J. (1998). A comparative review of the pharmacokinetics of boric acid in rodents and humans. *Biol. Trace Elem. Res., 66*, 331-341.

Naderi, A.S., Palmer, B.F. (2006). Successful treatment of a rare case of boric acid overdose with hemodialysis. *Am. J. Kidney dis., 48*, e95-e97.

Nagai, N., Murao, T., Okamoto, N., Ito, Y. (2010). Comparison of corneal wound healing rates after instillation of commercially available latanoprost and travoprost in rat debrided corneal epithelium. *J. Oleo. Sci., 59*, 135-141.

Nichols, J.J., Sinnott, L.T., King-Smith, P.E., Nagai, H., Tanikawa, S. (2008). Hydrogel contact lens binding induced by contact lens rewetting drops. *Optom. Vis. Sci.*, *85*, 236-240.

Nozawa, A., Miwa, K., Kobayashi, M., Fujiwara, T. (2006b). Isolation of *Arabidopsis thaliana* cDNAs that confer yeast boric acid tolerance. *Biosci. Biotechnol. Biochem.*, *70*, 1724-1730.

Nozawa, A., Takano, J., Kobayashi, M., von Wirén, N., Fujiwara, T. (2006a). Roles of BOR1, DUR3, and FPS1 in boron transport and tolerance in *Saccharomyces cerevisiae*. *FEMS Microbiol. Lett.*, *262*, 216-222.

Nyirjesy, P., Alexander, A.B., Weitz, M.V. (2005). Vaginal *Candida parapsilosis*: pathogen or bystander? *Infect. Dis. Obstet. Gynecol.*, *2005*, 37-41.

O'Neil, M.J., Ed. (2001). *The Merck Index*, 13th ed., Rahway, NJ: Merck and Co., Electronic Version.

Oliveira, D.P., Gomes, B.P., Zaia, A.A., Souza-Filho, F.J., Ferraz, C.C. (2008). Ex vivo antimicrobial activity of several bleaching agents used during the walking bleach technique. *Int. Endod. J.*, *41*, 1054-1058.

Ono, K., Kinashi, Y., Masunaga, S., Suzuki, M., Takagaki, M. (1998a). Effect of electroporation on cell killing by boron neutron capture therapy using borocaptate sodium (^{10}B-BSH). *Jpn. J. Cancer Res.*, *89*, 1352-1357.

Ono, K., Kinashi, Y., Masunaga, S., Suzuki, M., Takagaki, M. (1998b). Electroporation increases the effect of borocaptate (^{10}B-BSH) in neutron capture therapy. *Int. J. Radiat. Oncol. Biol. Phys.*, *42*, 823-826.

Otero, L., Fleites, A., Méndez, F.J., Palacio, V., Vázquez, F. (1999). Susceptibility of Candida species isolated from female prostitutes with vulvovaginitis to antifungal agents and boric acid. *Eur. J. Clin. Microbiol. Infect. Dis.*, *18*, 59-61.

Otero, L., Palacio, V., Mendez, F.J., Vazquez, F. (2002). Boric acid susceptibility testing of non-C. albicans Candida and *Saccharomyces cerevisiae*: comparison of three methods. *Med. Mycol.*, *40*, 319-322.

Ozcan, K.M., Ozcan, M., Karaarslan, A., Karaarslan, F. (2003). Otomycosis in Turkey: predisposing factors, aetiology and therapy. *J. Laryngol. Otol.*, *117*, 39-42.

Oztürkcan, S., Dündar, R., Katilmis, H., Ilknur, A.E., Aktas, S., Haciömeroğlu, S. (2009). The ototoxic effect of boric acid solutions applied into the middle ear of guinea pigs. *Eur. Arch. Otorhinolaryngol.*, *266*, 663-667.

Pandit, N. (1986). Chloramphenicol monograph, In: K.A. Connors, G.L. Amidon, V.J. Stella, Eds., *Chemical Stability of Pharmaceuticals: A*

Handbook for Pharmacists, 2nd ed., New York, NY: John Wiley & Sons, and references therein, pp. 328-335.

Patel, A., Agrawal, S.C. (2002). Sensitivity of ornithophillic fungi to some drugs. *Hindustan Antibiot. Bull.*, *44*, 49-52.

Pawelczyk, E., Matlak, B. (1982). Kinetics of drug decomposition. Part 74. Kinetics of degradation of minocycline in aqueous solution. *Pol. J. Pharmacol. Pharm.*, *34*, 409-421.

Piemjai, M., Surakompontorn, J. (2006). Effect of tooth-bleaching on the tensile strength and staining by caries detector solution on bovine enamel and dentin. *Am. J. Dent.*, *19*, 387-392.

Pinto, J.T., Rivlin, R.S. (1987). Drugs that promote renal excretion of riboflavin. *Drug Nutr. Interact.*, *5*, 143-151.

Pinto, J.T., Huang, Y.P., McConnel, R.J., Rivlin, R.S. (1978). Increased urinary riboflavin excretion resulting from boric acid ingestion. *J. Lab. Clin. Med.*, *92*, 126-134.

Pisani, M.X., Silva-Lovato, C.H., Malheiros-Segundo, Ade, L., Macedo, A.P., Paranhos, H.F. (2009). Bond strength and degree of infiltration between acrylic resin denture liner after immersion in effervescent denture cleanser. *J. Prosthodont.*, *18*, 123-129.

Plotino, G., Buono, L., Grande, N.M., Pameijer, C.H., Somma, F. (2008). Nonvital tooth bleaching: a review of the literature and clinical procedures. *J. Endod.*, *34*, 394-407.

Poller, F., Bauch, T., Sauerwein, W., Bocker, W., Wittig, A., Streffer, C. (1996). Comet assay study of DNA damage and repair of tumor cells following boron neutron capture irradiation with fast d(14) + Be neutrons. *Int. J. Radiat. Biol.*, *70*, 593-602.

Porter, I.A., Brodie, J. (1969). Boric acid preservation of urine samples. *Br. Med. J.*, *2*, 353-355.

Prentice, L.H., Tyas, M.J., Burrow, M.F. (2006). The effects of boric acid and phosphoric acid on the comprehensive strength of glass-ionomer cements. *Dent. Mat.*, *22*, 94-97.

Prutting, S.M., Cerveny, J.D. (1998). Boric acid vaginal suppositories: A brief review. *Infect. Dis. Obstet. Gynecol.*, *6*, 191-194.

Ray, D., Goswani, R., Banerjee, U., Dadhwal, V., Goswami, D., Mandal, P., Sreenivas, V., Kochupillai, N. (2007b). Prevalence of *Candida glabrata* and its response to boric acid vaginal suppositories in comparison with oral fluconazol in patients with diabetes and vulvovaginal candidiasis. *Diabetes Care, 30,* 312-317.

Ray, D., Goswani, R., Dadhwal, V., Goswami, D., Banerjee, U., Kochupillai, N. (2007a). Prolonged (3-month) mycological cure after boric acid suppositories in diabetic women with vulvovaginal candidiasis. *J. Infect.*, *55*, 374-377.

Reichman, O., Akins, R., Sobel, J.D. (2009). Boric acid addition to suppressive antimicrobial therapy for recurrent bacterial vaginosis. *Sex Transm. Dis.*, *36*, 732-734.

Reilly, W.J. Jr. (2006). Pharmaceutical necessities, In: R. Hendrickson, Ed., *Remington: The Science and Practice of Pharmacy*, 21st ed., Philadelphia, PA: Lippincott Williams & Wilkins, pp. 1083-1084, 1089.

Rivlin, R.S. (2007). Riboflavin (Vitamin B_2), In: J. Zemplini, R.B. Rucker, D.B. McCormick, J.W. Suttie, Eds., *Handbook of Vitamins*, 4th ed., Boca Raton, FL: CRC Press, p. 241.

Robson, H., Craig, D.Q., Deutsch, D. (2000). An investigation into the release of cefuroxime axetil from taste-masked stearic acid microspheres. II. The effects of buffer composition on drug release. *Int. J. Pharm.*, *195*, 137-145.

Romano, L., Battaglia, F., Masucci, L., Sanguinetti, M., Posteraro, B., Plotti, G., Zanetti, S., Fadda, G. (2005). In vitro activity of bergamot natural essence and furocoumarin-free and distilled extracts and their associations with boric acid, against clinical yeast isolates. *J. Antimicrob. Chemother.*, *55*, 110-114.

Rowe, R.C., Sheskey, P.J., Quinn, M.E. (2009). *Handbook of Pharmaceutical Excipients*, 6th ed., London: Pharmaceutical Press, Electronic Version.

Ruth, H.J., Eger, K., Troschutz, R. (1991). *Pharmaceutical Chemistry: Drug Analysis*, Vol. 2, London: Ellis Harwood, pp. 654-658.

Salphale, P.S., Shenoi, S.D. (2003). Contact sensitivity to calcium hypochlorite. *Contact Dermatitis*, *48*, 162.

Sassetti, R.J., Fudenberg, H.H. (1971). Alpha-methyldopa melanin. Synthesis and stabilization in vitro. *Biochem. Pharmacol.*, *20*, 57-66.

Schmid, R.W. (1970). Hydrolysis equilibrium of phenylbutazone (Butazolidin). *Helv. Chim. Acta*, *53*, 2239-2251.

Schon, J.S., Jansen, J.A., Aggerbeck, B. (1984). Human pharmacokinetics and safety of boric acid. *Arch. Toxicol. Suppl.*, *7*, 232-235.

Scorei, R., Cimpoiasu, V.M. (2006). Boron enhances the thermostability of carbohydrates. *Orig. Life Evol. Biosph.*. *36*, 1-11.

Scorei, R., Ciubar, R., Ciofrangeanu, C.M., Mitran, V., Cimpean, A., Iordachescu, D. (2008). Comparative effects of boric acid and calcium fructoborate on breast cancer cells. *Biol. Trace Elem. Res.*, *122*, 197-205.

Shaposhnikov, V.I., Zorik, V.V. (2001). Combined treatment of purulent-necrotic lesions of lower extremities in diabetic patients. *Khirurgiia (Mosk.)*, *2*, 46-49.

Sheng, M.H., Taper, L.J., Veit, H., Qian, H., Ritchey, S.J., Lau, K.H. (2001). Dietary boron supplementation enhanced the action of estrogen, but not that of parathyroid hormone, to improve trabecular bone quality in ovariectomized rats. *Biol. Trace Elem. Res.*, *82*, 109-123.

Shinohara, M.S., Peris, A.R., Pimenta, L.A., Ambrosano, G.M. (2005). Shear bond strength evaluation of composite resin on enamel and dentin after nonvital bleaching. *J. Esthet. Restor. Dent.*, *17*, 22-29.

Shinohara, M.S., Peris, A.R., Rodrigues, J.A., Pimenta, L.A., Ambrosano, G.M. (2004). The effect of nonvital bleaching on the shear bond strength of composite resin using three adhesive systems. *J. Adhes. Dent.*, *6*, 205-209.

Singh, S., Sobel, J.D., Bhargava, P., Boikov, D., Vazquez, J.A. (2002). Vaginitis due to *Candida krusei*: epidemiology, clinical aspects, and therapy. *Clin. Infect. Dis.*, *35*, 1066-1070.

Sinko, P.J. (2006). *Martins Physical Pharmacy and Pharmaceutical Sciences*, 5th ed., Philadelphia, PA: Lippincott Williams & Wilkins, pp. 416-420.

Sobel, J.D., Chaim, W. (1997). Treatment of *Torulopsis glabrata* vaginitis: retrospective review of boric acid therapy. *Clin. Infect. Dis.*, *24*, 649-652.

Sobel, J.D., Chaim, W., Nagappan, V., Leaman, D. (2003). Treatment of vaginitis caused by *Candida glabrata*: use of topical boric acid and flucytosine. *Am. J. Obstet. Gynecol.*, *189*, 1297-1300.

Soine, T.O., Wilson, C.O. (1967). *Rogers Inorganic Pharmaceutical Chemistry*, 8th ed., Philadelphia, PA: Lea & Febiger, pp. 125-126.

Sood, G., Nyirjesy, P., Weitz, M.V., Chatwani, A. (2000). Terconazole cream for non-*Candida albicans* fungal vaginitis: results of a retrospective analysis. *Infect. Dis. Obstet. Gynecol.*, *8*, 240-243.

Sponer, J.E., Sumpter, B.G., Leszczynski, J., Sponer, J., Fuentes-Cabrera, M. (2008). Theoretical study on the factors controlling the stability of the borate complexes of ribose, arabinose, lyxose and xylose. *Chemistry*, *14*, 9990-9998.

Sweetman, S.C., Ed. (2009). Boric acid, *Martindale: The Complete Drug Reference*, 35th ed., London: Pharmaceutical Press, Electronic Version.

Teixeira, E.C., Hara, A.T., Turssi, C.P., Serra, M.C. (2003). Effect of non-vital tooth bleaching on microleakage of coronal access restorations. *J. Oral Rehabil.*, *30*, 1123-1127.

Teixeira, E.C., Turssi, C.P., Hara, A.T., Serra, M.C. (2004). Influence of post-bleaching time intervals on dentin bond strength. *Braz. Oral Res.*, *18*, 75-79.

Teshima, D., Morishita, K., Ueda, Y., Futagami, K., Higuchi, S., Komoda, T., Nanishi, F., Taniyama, T., Yoshitake, J., Aoyema, T. (1992). Clinical management of boric acid ingestion: pharmacokinetic assessment of efficacy of hemodialysis for treatment of acute boric acid poisoning. *J. Pharmacobiodyn.*, *5*, 287-294.

Thierauf, A., Serr, A., Halter, C.C., Al-Ahmad, A., Rana, S., Weinmann, W. (2008). Influence of preservatives on the stability of ethyl glucuronide and ethyl sulphate in urine. *Forensic. Sci. Int.*, *182*, 41-45.

Thongboonkerd, V., Saetun, P. (2007). Bacterial overgrowth affects urinary proteome analysis: recommendation for centrifugation, temperature, duration, and the use of preservatives during sample collection. *J. Proteome Res.*, *6*, 4173-4181.

Thue, G., Baerheim, A., Bjelkarøy, W.I., Digranes, A. (2010). Dip-slides for culturing urine in general practice. *Tidsskr Nor Laegeforen*, *130*, 483-486.

Timpawat, S., Nipattamanon, C., Kijsamanmith, K., Messer, H.H. (2005). Effect of bleaching agents on bonding to pulp chamber dentine. *Int. Endod. J.*, *38*, 211-217.

Ting, R., Harwig, C., auf dem Keller, U., McCormick, S., Austin, P., Overall, C.M., Adam, M.J., Ruth, T.J., Perrin, D.M. (2008). Toward [18F]-labeled aryltrifluoroborate radiotracers: in vivo positron emission tomography imaging of stable aryltrifluoroborate clearance in mice. *J. Am. Chem. Soc.*, *130*, 12045-12055.

United States Pharmacopeia 30 / National Formulary 25 (2007). Rockville, MD: United States Pharmacopeial Convention, Electronic Version.

Usuda, K., Kono, K., Orita, Y., Dote, T., Iguchi, K., Nishiura, H., Tominaga, M., Tagawa, T., Goto, E., Shirai, Y. (1998). Serum and urinary boron levels in rats after single administration of sodium tetraborate. *Arch. Toxicol.*, *72*, 468-474.

Valera, M.C., Camargo, C.H., Carvalho, C.A., de Oliveira, L.D., Camargo, S.E., Rodrigues, C.M. (2009). Effectiveness of carbamide peroxide and sodium perborate in non-vital discolored teeth. *J. Appl. Oral Sci.*, *17*, 254-261.

VanKessel, K., Assefi, N., Marrazzo, J., Eckhert, L. (2003). Common complementary and alternative therapies for yeast vaginitis and bacterial vaginosis: a systematic study. *Obstet. Gynecol. Surv.*, *58*, 351-358.

Van Slyke, K.K., Michel, V.P., Rein, M.F. (1981). Treatment of vulvovaginal candidiasis with boric acid powder. *Am. J. Obstet. Gynecol., 141*, 145-148.

Vázquez, F., Otero, L., Ordás, J., Junquera, M.L., Varela, J.A. (2004). Up to date in sexually transmitted infections: epidemiology, diagnostic approaches and treatments. *Enferm. Infecc. Microbiol. Clin., 22*, 392-411.

Wadke, D. A., Guttman, D. E. (1964). Complex formation influence on reaction rate. II. Hydrolytic behaviour of riboflavin in borate buffer. *J. Pharm. Sci. 53*, 1073-1076.

Walmod, P.S., Gravemann, U., Nau, H., Berezin, V., Bock, E. (2004). Discriminative power of an assay for automated in vitro screening of teratogens. *Toxicol. In Vitro, 18*, 511-525.

Wang, C., Armstrong, D.W., Risley, D.S. (2007). Empirical observations and mechanistic insights on the first boron-containing chiral selector for LC and supercritical fluid chromatography. *Anal. Chem., 79*, 8125-8135.

Watson, B.L., Cormier, R.A., Harbeck, R.J. (1998). Effect of pH on the stability of methacholine chloride in solution. *Respir. Med., 92*, 588-592.

Wiegand, A., Drebenstedt, S., Roos, M., Magalhães, A.C., Attin, T. (2008). 12-month color stability of enamel, dentine, and enamel-dentine samples after bleaching. *Clin. Oral Investig., 12*, 303-310.

Wilson, A.D., Kent, B.E. (1972). A new translucent cement in dentistry. The glass ionomer cement. *Br. Dent. J., 132*, 133-135.

Wood, R.M. (1986). Carbenicillin monograph, In: K.A. Connors, G.L. Amidon, V.J. Stella, Eds., *Chemical Stability of Pharmaceuticals: A Handbook for Pharmacists*, 2nd ed., New York, NY: John Wiley & Sons, and references therein, pp. 290-294.

Wood, W.G. (2009). The determination of uric acid in urine--forgotten problems rediscovered in an external quality assessment scheme. *Clin. Lab., 55*, 341-352.

Yamana, T., Tsuji, A. (1976). Comparative stability of cephlosporins in aqueous solution: Kinetics and mechanisms of degradation. *J. Pharm. Sci., 65*, 1563-1574.

Yelvigi, M. (2005). Boric acid, In: R.C. Rowe, P.J. Sheskey, S.C. Owen, Eds., *Pharmaceutical Excipients*, 5th ed., London: Pharmaceutical Press, Electronic Version.

Yoshida, M., Tokiyasu, T., Watabiki, T., Ueda, M., Ishida, N. (1991). Study on the histochemical staining of boric acid. *Nihon Hoigaku Zasshi, 45*, 416-422.

Yoshida, M., Watabiki, T., Tokiyasu, T., Ishida, N. (1989a). Determination of boric acid in biological materials by curcuma paper. *Nihon Hoigaku Zasshi*, *43*, 497-501.

Yui, K.C., Rodrigues, J.R., Mancini, M.N., Balducci, I., Gonçalves, S.E. (2008). Ex vivo evaluation of the effectiveness of bleaching agents on the shade alteration of blood-stained teeth. *Int. Endod. J.*, *41*, 485-492.

Zvirblis, P., Socholitsky, I., Kondritzer, A. (1956). The kinetics of the hydrolysis of atropine. *J. Pharm. Sci.*, *45*, 450-454.

Chapter 4

PHARMACOLOGICAL ASPECTS OF BORATES

4.1. INTRODUCTION

Humans consume daily about a milligram of boron, mostly from fruits and vegetables. At high doses, boron is a developmental and reproductive toxin in animals. Prolonged exposure to pharmacologically active levels of boric acid, the naturally occurring form of boron in human plasma, induces morphological changes in cells. Boric acid is a component of several consumer goods and is used in numerous industrial processes. The potential widespread human exposure to borates may cause hypotension, metabolic acidosis, oliguria, chemesthesis and other symptoms. The toxicity, absorption, metabolism, oxidative stress and related aspects of borates are presented in the following sections:

4.2. TOXICITY OF BORATES

4.2.1. Human Toxicity

The improper use of boric acid containing antiseptics is one of the most common causes of toxic accidents in newborns and infants. Health hazards may also arise from inadvertent absorption of insecticides and household products containing borates as well as from occupational accidents related to production and use of borates. Most of the boronated compounds with hypolipidemic, antiinflammatory or anticancer properties have been found to be highly toxic at the required therapeutic dosages in animals (Locatelli *et al.*,

1987; Craan *et al.*, 1997). The absorption of boric acid from damaged skin may reach toxic levels and cause fatal poisoning, particularly in infants, on topical application to burns, denuded areas, granulation tissues and serous cavities. It may cause serious poisoning from oral ingestion of as little as 5 g. Symptoms of poisoning include nausea, vomiting, abdominal pain, diarrhea, headache, depression, alopecia and visual disturbance. The kidney may be injured resulting in death. Boric acid accumulates principally in the brain, liver and body fat, with the greatest amount in the brain. It is the grey matter of the cerebrum and the spinal cord with the greatest content of boron following death (Soine and Wilson, 1967; Browning, 1969; Gosselin *et al.*, 1976; Seigel and Wilson, 1986; Linden *et al.*, 1986; Litovitz *et al.*, 1988; Kiesche-Nesselrodt and Hooser, 1990; Werfel *et al.*, 1998; Teshima *et al.*, 2001; Beckett *et al.*, 2001; Pahl *et al.*, 2005; Reilly, 2006; Cain *et al.*, 2008; Ahmad *et al.*, 2010a,b).

Accidental boric acid poisoning by food containing boric acid as a preservative (Chao *et al.*, 1991; Tangermann *et al.*, 1992) and by ingestion of a boric acid containing pesticide by children (Hamilton and Wolf, 2007) or boric acid powder by an old person (Corradi *et al.*, 2010) has been reported. Several cases of acute ingestions of boric acid resulting in death have occurred (Restuccio *et al.*, 1992; Ishii *et al.*, 1993; Matsuda *et al.*, 2004). Human exposure to boron and its compounds can arise from a variety of natural sources, such as soil and water, or through its use in pesticides, preservatives, pharmaceuticals and household products (Siegel and Wilson, 1986; Woods, 1994; Moore, 1997; Richold, 1998; Coughlin, 1998; Hubbard, 1998; Cain *et al.*, 2008) and may be hazardous to health.

4.2.2. Embryotoxicity

It has been observed that fishes in the embryo-larval stage of development are sensitive to boron (0-500 μM) as boric acid. Chronic exposure below 9 μM B/L impairs embryonic growth and above 10 mM B/L causes death (Rowe *et al.*, 1998). Boric acid induces dysmorphogenesis in the skull, vertebral column and ribs in both human population and laboratory animals (Tyl *et al.*, 2007).

Prenatal exposure of rats to elevated levels of boric acid (500 mg/kg) causes embryonic malformations of the axial skeleton involving the head, sternum, ribs and vertebrae (Narotsky *et al.*, 1998). Boric acid (500-750 mg/kg), administered to pregnant CD-1 mice once daily on gestational days 6-10 has been shown to produce ribogenesis, a rare effect in laboratory animals.

It has been suggested that boric acid affects early processes such as gastrulation and presomitic mesoderm formation (Cherrington and Chernoff, 2002). Axial skeletal defects in boric acid-exposed rat embryos are associated with anterior shifts of hox gene expression domains (Wery et al., 2003). The skeletal defects resulting from combined exposure to hyperthermia and boric acid are additive for segmentation defects and synergistic for the reduction in numbers of vertebrae (Harrouk et al., 2005). Boric acid inhibits embryonic histone deacetylases in mouse suggesting a molecular mechanism for the induction of skeletal malformations and boric acid-related teratogenicity (Di Renzo et al., 2007). The embryonic stem cell test (EST) has been developed as a ECVAM-validated assay to detect embryotoxicity (Peters et al., 2008a). The automated image recording of contractile cardiomyocyte-like cells in the EST allows for an unbiased high throughput method to assess the relative embryotoxic potency of some test compounds. The values obtained for these compounds are: 6-aminonicotinamide (1) > valproic acid (0.007-0.013) > boric acid (0.002-0.005) > penicillin G (0.00001) (Peters et al., 2008b).

Blood boron concentrations of 1.27 ± 0.298 and 1.53 ± 0.546 µg B/g are associated with the no-observed-adverse effect level (NOAEL) (10 mg/B/kg/d) and lowest-observed-adverse-effect level (13 mg B/kg/d) with developmental toxicity in pregnant rats fed boric acid throughout gestation (Price et al., 1997).

4.2.3. Developmental and Reproductive Toxicity

Boron is essential for humans and the daily requirement is about 1 mg. At high doses, boron is a developmental and reproductive toxin in animals. An oral NOAEL of 9.6 mg B/kg/d has been established for developmental toxicity in Sprague-Dawley rats fed boric acid (Pahl et al., 2001). Boric acid and inorganic borates cause both the developmental and reproductive toxicity in mice (Harris et al., 1992), rats (Chapin et al., 1997, 1998; Fail et al., 1998; Wang et al., 2008), and rabbits (Price et al., 1996a).

Several studies have been conducted on the developmental toxicity of boric acid in mice, rats and rabbit (Heindel et al. 1992, 1994; Price et al., 1996b, 1998; Vaziri et al., 2001). The developmental toxicity NOAEL in the rat is 0.075% boric acid (55 mg/kg/d) on gestational day 20 and 0.1% boric acid (74 mg/kg/d) on postnatal day 21 (Price et al., 1996b). A human risk assessment study has been conducted to derive an appropriate safe exposure level of boric acid and borax in drinking water. A rat developmental toxicity study with a NOAEL of 9.6 mg B/kg/d was selected as a basis for the risk assessment since it represents the most sensitive end point of toxicity. The results showed that boron in US drinking water (0.031 B/L) is much below the toxic level (2.44 mg B/L) and does not pose any health risk to the public (Murray, 1995).The benchmark dose (BMD) approach has been proposed as an alternative basis for reference value calculations in the assessment of developmental toxicity (Allen et al., 1996).

The potential reproductive toxicity of boric acid in Swiss mice has been evaluated using the Reproductive Assessment by Continuous Breeding (RACB) protocol. The results establish the reproductive toxicity of boric acid in mice and demonstrate that the male is the most sensitive sex (Fail et al., 1991). Acute oral exposure to boric acid (2000 mg/kg) adversely affects spermiation and sperm quality in the male rat. The no effect level is 500 mg/kg (Linder et al., 1990). Development of testicular lesions in adult rats after treatment with high dose boric acid (6000-9000 ppm) has been observed. This is characterized by decreased fertility, sperm motility and inhibited spermiation followed by atrophy (Treinen and Chapin, 1991; Ku et al., 1993a,b; Chapin and Ku, 1994; Yoshizaki et al., 1999). A mechanism of the testicular toxicity of boric acid based on the involvement of serine proteases plasminogen activators in spermiation has been proposed (Ku and Chapin, 1994). Boric acid feed (4500-9000 ppm) in mice is a potent reproductive toxicant in males and females (Fail et al., 1991). Testicular effects occur at 26 mg B/kg/d. The endocrine toxicity of boronated compounds includes altered follicle stimulating hormone and testosterone within 14 days of treatment. The most sensitive of all the end points are prenatal growth and morphologic development in the rat. These changes occurred at a dose of 12.9 mg B/kg/d and the NOAEL for rat fetal development was 9.6 mg B/kg/d (Fail et al., 1998).

Collaborative studies to evaluate toxicity on male reproductive system by repeated dose studies in rats have shown that at the appropriate dose (500 mg/kg), the testicular toxicity of boric acid can be detected after only two weeks of repeated daily oral treatment (Kudo et al., 2000; Fukuda et al., 2000). Boric acid has been evaluated for reproductive toxicity in Xenopus laevis. This model appears to be a useful tool in the initial assessment of potential reproductive toxicants for further testing (Fort et al., 2001). Among NTP reproductive toxicants, boric acid has been identified as high priority for occupational studies to determine safe versus adverse reproductive effects (Robbins et al., 2009).

X-ray micro-computed tomography has been applied to the assessment of oric acid-induced fetal skeletal toxicity in Sprague-Dawly rats. It produces high-resolution images of skeletal structures and can be used as an alternative method to alizarin red staining (Wise and Winkelmann, 2009).

4.2.4. Genotoxicity

The genotoxic potential of boric acid in *Escherichia coli* PQ 37 has been assessed in the presence of aflatoxin B1 using SOS chromotest. Boric acid induces beta-galactosidase synthesis on the tester bacteria both in the presence and absence of S9 activation mixure. It may possibly act as a syngenotoxic and/or a cogenotoxic agent (Odunola, 1997). The effect of boric acid as a food preservative on root tips of *Allium cepa* L has been studied. Boric acid (20 to 100 ppm) reduced mitotic division in *A. cepa* compared with that of the respective control. Mitotic index values were generally decreased with increasing concentrations of boric acid and the longer period of treatment (5 to 20 h) (Turkoglu, 2007). The genotoxic effects of boric acid have been determined using sister chromatid and chromosome aberration tests in human peripheral lymphocytes. Boric acid induced structural and total chromosome aberration at 400–1000 μg / mL (Arsalan *et al.*, 2008). The genetic effects of boric acid and borax (2.5 to 10 μM) on human whole blood cultures with and without the addition of titanium oxide (a carcinogenic / mutagenic agent) have been investigated. Both compounds have been found to cause increased resistance of DNA to damage induced by titanium oxide (Turkez, 2008). DNA alterations in bean seedlings offer useful biomarker assay for the evaluation of genotoxic effects of boron pollution in plants (Cenkci *et al.*, 2009). Borax (0.15–0.3 mg / mL) affects the cell and human chromosomes (both numerical and structural abnormalities) and may lead to genetic defects (Pongsavee, 2009). Boric acid has been found to prevent the genotoxicity of paclitaxel, an anticancer drug used for the treatment of breast, ovarian and lung cancers (Turkez *et al.*, 2010).

4.2.5. Phytotoxicity

A phytotoxicity test has been applied to evaluate the toxicity of toxicants in Arctic soils. The phytotoxicity of boric acid in cryoturbated soils of polar region is much greater than that in other soils. A boric acid concentration of less than 150 μg/g soil is needed to inhibit root and shoot growth by 20%. Northern wheat grass (*Elymus lanceolatus*) is more sensitive to toxicants in

Arctic soil than the other plants (Anaka *et al.*, 2008a). Arctic soils are more resistant to ammonium nitrate toxicity compared with that of the temperate soils under specific conditions (Anaka *et al.*, 2008b).

4.3. ABSORPTION OF BORIC ACID

Three percent boric acid incorporated in an anhydrous, water emulsifying ointment causes no increase of boron levels in blood and urine during a period of 1–9 days after a single topical application. The same amount of boric acid incorporated in a water-based jelly does cause an increase in blood and urine levels, beginning within 2-6 h after application. The decisive factor is the degree of liberation of boron depending on the type of vehicle (Stuttgen *et al.*, 1982). Boric acid taken orally by six male volunteers is absorbed to equal extents from a 3% water solution and an anhydrous, water emulsifying ointment. Virtually complete gastro-intestinal absorption and renal excretion are indicated by the 96 h urinary recovery amounting to 89.1–98.3% and 89.2–97.5%, of the dose ingested as solution and ointment, respectively. This indicates that the formulation of the ointment is an important factor in determining the release of boric acid on external application but it does not alter the absorption of boric acid on ingestion of the ointment (Jansen *et al.*, 1984a). The absorption of boric acid through damaged skin from a body ointment has been measured by an in vitro permeation method. The maximum permeation was 14.1 $\mu g/cm^2$ of boric acid, which corresponds to 86% of the release from the ointment. The results indicate that the degree of permeation through damaged skin depends on the degree of liberation from the vehicle (Dusemund, 1987). The release of biocides (benzoic, sorbic and boric acids) incorporated into modified silica films has been investigated with respect to composite structure. The liberation rates of the embedded acids are proportional to the biocide-to-silica ratio and are changed by adding soluble polymers such as hydroxypropylcellulose (Bottcher *et al.*, 1999).

Dordas and Brown (2001) have suggested boron (in the form of undissociated boric acid) as an essential micronutrient for animals. They determined the permeability coefficient of boric acid in frog (*Xenopus oocytes*) as 1.5×10^{-6} cm/s, which is very close to the permeability across liposomes made with phosphatidylcholine and cholesterol (the major lipids in the oocyte membrane). The corneal penetration of CS-088, an ophthalmic agent, is significantly enhanced in the presence of EDTA / boric acid by approximately 1.6 fold. The permeability-enhancing effect of EDTA / boric acid is apparently synergistic and concentration dependent on both EDTA and boric acid (Kikuchi *et al*., 2005a). In the corneal permeability of CS-088 (4-[1-hydroxy-1-methylethyl]-2-propyl-1-[4-[2-[tetrazole-5-yl] phenyl] methylimidazole-5-carboxylic acid monohydrate), EDTA / boric acid significantly increased the membrane fluidity of liposomes (Kikuchi *et al*., 2005b). The dermal route is important in exposure to biocidal products including boric acid. Repeated exposure of these products (specially undiluted) increases skin permeability causing harmful effects (Buist *et al*., 2005).

4.4. METABOLISM

Boric acid plays an important role in bone development. Dietry boron (in the form of boric acid) may be beneficial for optimum calcium metabolism and as a consequence, optimal bone metabolism. Serum levels of minerals as well as osteocalcin (a marker of bone resorption) are dependent to a greater extent on the hormonal (17 beta-estradiol) status of the animal. Boron treatment increases the hormonal-induced elevation of urinary calcium and magnesium. The bone mineral density of the L5 vertebra and proximal femur is highest after hormonal treatment (Gallardo-Williams *et al*., 2003).

A boric acid solution (275 mg/L), administered to rats for 30 days orally in an "ad libitum" dose, has been found to cause changes in the lipid metabolism, in the water-mineral balance of the organism and in the acid-base equilibrium of the blood. The boric acid solution, suitably diluted with a nutritive fluid, acted spastically on the peristalsis of the small intestine of the rabbit (Drobnik and Latour, 2001).

The influence of dietary boron supplementation on some serum parameters and egg-yolk cholesterol has been studied in laying hens. Serum gamma-glutamyl transpeptidase (GGT) activity, albumin, glucose, total cholesterol and HDL– and LDL–cholesterol levels are decreased with all boron levels (10-400 mg/kg boric acid) (Eren and Uyanik, 2007). An investigation of the dietary effects of boric acid (156-2500 ppm) on the protein profiles in greater wax moth (*Galleria mellonella*) has been carried out. Many undetermined protein fractions (6.5-260 kDa), in addition to well-defined protein fractions such as lipophorins and storage proteins, have been detected in the tissues. A marked quantitative change in the 45kDa protein fraction of the hemolymph has been observed in the VIIth instar larvae reared on 2500 ppm boric acid (Hyrsl *et al*., 2008). The presence of boron has been found to decrease the synthesis of proteoglycan, collagen and total proteins in the pelvic cartilage of chick embryo (Benderdour et al., 1997).

4.5. OXIDATIVE STRESS

Oxidative stress plays an important role during inflammatory diseases. Boric acid has been implicated to modulate certain inflammatory mediators and regulate inflammatory processes. Boric acid could inhibit the lipopolysaccharide-induced formation of tumor necrosis factor-alpha and ameliorate the d,l-buthionine-S,R-sulfoximine induced glutathione depletion in TNF-1 cells (Cao et al., 2008).

Peripheral blood cultures exposed to various doses (5-500 mg/L) of boron compounds have been tested for DNA damage and oxidative stress. Boron compounds at low doses are useful in supporting antioxidant enzyme activities in human blood cultures, though in increasing doses they constitute oxidative stress (Turkez et al., 2007). The effects of boric acid induced oxidative stress on antioxidant enzymes and survivorship of wax moth larvae (Galleria mellonella) have been studied. The content of malondialdehyde (an oxidative stress indicator) was found to be significantly increased indicating that boric acid toxicity is related, in part, to oxidative stress management (Hyrsl et al., 2007).

The oxidative stress and antioxidant responses to boron (as boric acid) of chickpea plants differing in their tolerance to drought have been investigated. In comparison to the control group, all enzyme activities (except ascorbate peroxidase and superoxide dismutase) decrease with 1.6 mM boron stress. The glutathione reductase activity decreased, while activities of catalase, peroxidase and ascorbate peroxidase did not change with 6.44 mM boron (Ardic et al., 2009).

4.6. ENZYME INHIBITION

Various enzymes have been found to be reversibly or irreversibly inhibited by boric acid (a competitive inhibitor). It prevents jack bean urease from irreversible pyrocatechol (Kot and Zaborska, 2003) and allicin inactivation (Juszkiewicz et al., 2003). Urease from the seeds of pigeon pea is competitively inhibited by boric acid and boronic acids in the order of 4-

bromophenylboronic acid > boric acid > butylboronic acid > phenylboronic acid. Urease inhibition by boric acid is maximal at pH 5.0 and minimal at pH 10.0. The trigonal planar $B(OH)_3$ form is a more effective inhibitor than the tetrahedral $B(OH)_4^-$ anionic form (Reddy and Kayastha, 2006).

The addition of boric acid to an *in vitro* pre-mRNA splicing reaction causes a dose-dependent reversible inhibition effect on the second step of splicing. The mechanism of action does not involve chelation of metal ions, hindrance of 3' splice-site, or binding to h Slu7 (Shomron and Ast, 2003). Boric acid inhibits adenosine diphosphate-ribosyl cyclase (ADP-ribosyl cyclase) non-competitively. ADP-ribosyl cyclase converts NAD^+ to cyclic ADP-ribose (cADPR) and nicotinamide. Boric acid binding to cADPR is characterized by an apparent binding constant of 655 ± 99 L/mol at pH 10.3 (Kim *et al.*, 2006).

The activity of penicillin G acylase from *Alcaligenes faecalis* is increased 7.5 fold when cells are permeabilized with 0.3% (w/v) cetyl trimethyl ammonium bromide (CTAB). The treated cells are then entrapped by polyvinyl alcohol crosslinked with boric acid and glutaraldehyde to increase the stability. The conversion yield of penicillin G to 6-aminopenicillanic acid is 75% by the immobilized system (Cheng *et al.*, 2006).

4.7. SYNTHESIS OF BIOACTIVE COMPOUNDS

The 2-arachidonoylglycerol, an endogenous cannabinoid receptor ligand, has been synthesized from 1,3-benzylideneglycerol and arachidonic acid in the presence of N,N'-dicyclohexylcarbodiimide and 4-dimethylaminopyridine followed by treatment with boric acid and trimethyl borate. It stimulates NG108-15 cells to induce rapid transient elevation of the intracellular free Ca^{2+} concentrations through a CB1 receptor-dependent mechanism (Suhara *et al.*, 2000).

The synthesis of a polyglycerol with dendritic structure has been achieved by ring-opening polymerization of deprotonated glycidol. The polyglycerol is reacted with O-carboxymethylated chitosan dendrimer. The reaction of the dendrimer with boric acid results in a marked increase in the bulk viscosity indicating that boron can initiate the formation of a charge transfer complex. The complex shows significant activity against *Staphylococcus aureus* and *Pseudomonas aeruginosa* (Alencar de Queiroz *et al.*, 2006).

The action of boric acid at the molecular level has been examined using cell-free systems of transportation (isolated placenta nuclei) and translation (wheat germ extract). It has been found that 10 mM boric acid greatly increases RNA synthesis. Full length functional mRNA is produced because proteins of 14-80 kDa are translated (Dzondo-Gadet *et al.*, 2002). Incorporation of a biotinylated organometallic catalyst into a host protein (avidin or streptavidin) affords versatile artificial metalloenzymes for the reduction of ketones by transfer hydrogenation. The boric acid-formate mixture has been used as a hydrogen source compatible with these enzymes (Letondor *et al.*, 2005).

A highly stereoselective method for carbon-glycosidation of D-glucal with various potassium alkynyltrifluoroborates, involving oxonium intermediates, preferentially provides the alpha-acetyleneglycoside products with good yields (Vieira *et al.*, 2008). The 2-substituted 2H-chromene derivatives from salicylaldehyde using potassium vinylic borates in the presence of secondary amines have been synthesized. These compounds may prove useful to develop therapeutic agents (Torregroza *et al.*, 2009). A diol-SnCl$_4$ complex as chiral protic acid catalyst has been reported. It provides unpresidented levels of enantioselectivity in the catalytic allylation, methallyation, crotylation and 2-bromoallylation of aliphatic aldehydes. This catalytic allylboration methodology has been used for the efficient synthesis of the naturally occurring pyranone (+)-dodoneine and biologically important exomethylene-gamma-lactones (Rauniyar and Hall, 2009).

4.8. ANALYSIS IN BIOLOGICAL FLUIDS

Various analytical methods have been used to determine boric acid and borates in biological fluids. The determination of borates in biological fluids and tissues has been carried out by colorimetric methods using carminic acid as reagent. The technique of electrothermal atomic absorption spectrometry (ETAAS) has been applied to determine boron concentrations in blood in cases of acute poisoning (Moffat *et al.*, 2004). Boric acid in biological materials has been determined by its reaction with protonated curcumin and measurement of absorbance at 550 nm (Yoshida *et al.*, 1989). Quantitative determination of percutaneous absorption of ^{10}B in ^{10}B-enriched boric acid and borates in biological materials has been achieved by inductively coupled plasma-mass spectrometry (ICP-MS) (Wester *et al.*, 1998a,b,c). ICP-MS has been found to be highly sensitive for the measurement of boron concentrations in blood and tissues in patients undergoing boron neutron capture therapy of certain cancers (Morten and Delves, 1999).

Boric acid undergoes complex formation with a number of compounds (Saxena and Verma, 1983; Mendham *et al.*, 2000; Ahmad *et al.*, 2008, 2009; 2010a,b). This property has been used for the spectrophotometric determination of aminophylline in serum (Li *et al.*, 2008), fluorimetric determination of rubomycin in blood, milk and urine (Alykov *et al.*, 1976), amperometric determination of glucose in blood or plasma (Macholan *et al.*, 1992), potentiometric determination of zolpidem hemitartrate in biological fluids (Kelani, 2004), gas chromatographic-ion trap tandem-mass spectrometric determination of organic acids in urine (Pacenti *et al.*, 2008), capillary zone electrophoretic determination of ascorbic acid and uric acid in plasma (Zinellu *et al.*, 2004), icariin and its metabolites in serum (Liu and Lou, 2004), uric acid in urine (Zhao *et al.*, 2008), α- and β-amanitun in urine (Robinson-Fuentes *et al.*, 2008), cysteinylglycine in plasma (Zinellu *et al.*, 2008), bromide in serum samples (Pascali *et al.*, 2009) and microchip electrophoretic determination of glucose in blood (Maeda *et al.*, 2007).

REFERENCES

Ahmad, I., Ahmed, S., Sheraz, M.A., Vaid, F.H.M. (2008). Effect of borate buffer on the photolysis of riboflavin in aqueous solution. *J. Photochem. Photobiol. B: Biol.*, *93*, 82-87.

Ahmad, I., Ahmed, S., Sheraz, M.A., Vaid, F.H.M. (2009). Analytical applications of borates. *Mat. Sci. Res. J.*, *3*, 173-202.

Ahmad, I., Ahmed, S., Sheraz, M.A., Vaid, F.H.M. (2010a). Borate: toxicity, effect on drug stability and analytical applications, In: M.P. Chung, Ed., *Handbook on Borates: Chemistry, Production and Applications*, New York, NY: Nova Science Publishers Inc., Chap. 2.

Ahmad, I., Ahmed, S., Sheraz, M.A., Iqbal, K., Vaid, F.H.M. (2010b). Pharmacological aspects of borates. *Int. J. Med. Biol. Front.*, *16*, 977-1004.

Alencar de Queiroz, A.A., Abraham, G.A., Pires Camillo, M.A., Higa, O.Z., Silva, G.S., del Mar Fernández, M., San Román, J. (2006). Physicochemical and antimicrobial properties of boron-complexed polyglycerol-chitosan dendrimers. *J. Biomater. Sci. Polym. Ed.*, *17*, 689-707.

Allen, B.C., Strong, P.L., Price, C.J., Hubbard, S.A., Daston, G.P. (1996). Benchmark dose analysis of developmental toxicity in rats exposed to boric acid. *Fundam. Appl. Toxicol.*, *32*, 194-204.

Alykov, N.M., Karibiants, M.A., Alykova, T.V. (1976). Fluorimetric determination of rubomycin with the boric acid reaction. *Antibiotiki, 21,* 920-921.

Anaka, A., Wickstrom, M., Siciliano, S.D. (2008a). Increased sensitivity and variability of phytotoxic responses in Arctic soils to a reference toxicant, boric acid. *Environ. Toxicol. Chem., 27,* 720-726.

Anaka, A., Wickstrom, M., Siciliano, S.D. (2008b). Biogeochemical toxicity and phytotoxicity of nitrogenous compounds in a variety of arctic soils. *Environ. Toxicol. Chem., 27,* 1809-1816.

Ardic, M., Sekmen, A.H., Tokur, S., Ozdemir, F., Turkan, I. (2009). Antioxidant responses of chickpea plants subjected to boron toxicity. *Plant Biol. (Stuttg), 11,* 328-338.

Arslan, M., Topaktas, M., Rencuzogullari, E. (2008). The effects of boric acid on sister chromatid exchanges and chromosome aberrations in cultured human lymphocytes. *Cytotechnology, 56,* 91-96.

Beckett, W.S., Oskvig, R., Gaynor, M.E., Goldgeier, M.H. (2001). Association of reversible alopecia with occupational topical exposure to common borax-containing solutions. *J. Am. Acad. Dermatol., 44,* 599-602.

Benderdour, M., Hess, K., Gadet, M.D., Dousset, B., Nabet, P., Belleville, F. (1997). Effect of boric acid solution on cartilage metabolism. *Biochem. Biophys. Res. Commun., 234,* 263-268.

Böttcher, H., Jagota, C., Trepte, J., Kallies, K.H., Haufe, H. (1999). Sol-gel composite films with controlled release of biocides. *J. Control Release, 60,* 57-65.

Browning, E. (1969). *Toxicity of Industrial Metals,* 2nd ed., New York, NY: Appleton-Century-Crofts, pp. 90-97.

Buist, H.E., van de Sandt, J.J., van Burgsteden, J.A., de Heer, C. (2005). Effects of single and repeated exposure to biocidal active substances on the barrier function of the skin in vitro. *Regul. Toxicol. Pharmacol., 43,* 76-84.

Cain, W.S., Jalowayski, A.A., Schmidt, R., Kleinman, M., Magruder, K., Lee, K.C., Culver, B.D. (2008). Chemesthetic responses to airborne mineral dusts: boric acid compared to alkaline materials. *Int. Arch. Occup. Environ. Health, 81,* 337-345.

Cao, J., Jiang, L., Zhang, X., Yao, X., Geng, C., Xue, X., Zhong, L. (2008). Boric acid inhibits LPS-induced TNF-alpha formation through a thiol-dependent mechanism in THP-1 cells. *J. Trace Elem. Med. Biol., 22,* 189-195.

Cenkci, S., Yildiz, M., Cigerci, I.H., Konuk, M., Bozdag, A. (2009). Toxic chemicals-induced genotoxicity detected by random amplified polymorphic DNA (RAPD) in bean (*Phaseolus vulgaris* L.) seedlings. *Chemosphere*, 76, 900-906.

Chao, T.C., Maxwell, S.M., Wong, S.Y. (1991). An outbreak of aflatoxicosis and boric acid poisoning in Malaysia: a clinicopathological study. *J. Pathol.*, 164, 225-233.

Chapin, R.E., Ku, W.W. (1994). The reproductive toxicity of boric acid. *Environ. Health Perspect.*, 102 (Suppl.7), 87-91.

Chapin, R.E., Ku, W.W., Kenney, M.A., McCoy, H. (1998). The effects of dietary boric acid on bone strength in rats. *Biol. Trace Elem. Res.*, 66, 395-399.

Chapin, R.E., Ku, W.W., Kenney, M.A., McCoy, H., Gladen, B., Wine, R.N., Wilson, R., Elwell, M.R. (1997). The effects of dietary boron on bone strength in rats. *Fundam. Appl. Toxicol.*, 35, 205-215.

Cheng, S., Wei, D., Song, Q., Zhao, X. (2006). Immobilization of permeabilized whole cell penicillin G acylase from Alcaligenes faecalis using pore matrix crosslinked with glutaraldehyde. *Biotechnol. Lett.*, 28, 1129-1133.

Cherrington, J.W., Chernoff, N. (2002). Periods of vertebral column sensitivity to boric acid treatment in CD-1 mice in utero. *Reprod. Toxicol.*, 16, 237-243.

Corradi, F., Brusasco, C., Palermo, S., Belvederi, G. (2010). A case report of massive acute boric acid poisoning. *Eur. J. Emerg. Med.*, 17, 48-51.

Coughlin, J.R. (1998). Sources of human exposure: overview of water supplies as sources of boron. *Biol. Trace Elem. Res.*, 66, 87-100.

Craan, A.G., Myres, A.W., Green, D.W. (1997). Hazard assessment of boric acid in toys. *Regul. Toxicol. Pharmacol.*, 26, 277-280.

Di Renzo, F., Cappelletti, G., Broccia, M.L., Giavini, E., Menegola, E. (2007). Boric acid inhibits embryonic histone deacetylases: a suggested mechanism to explain boric acid-related teratogenicity. *Toxicol. Appl. Pharmacol.*, 220, 178-185.

Dordas, C., Brown, P.H. (2001). Permeability and the mechanism of transport of boric acid across the plasma membrane of *Xenopus laevis* oocytes. *Biol. Trace Elem. Res.*, 81, 127-139.

Drobnik, M., Latour, T. (2001). Investigation of the pharmacodynamic properties of the solution of boric acid. *Rocz. Panstw. Zakl. Hig.*, 52, 329-336.

Dusemund, B. (1987). Liberation and in vitro skin permeation of boric acid from an ointment. *Arzneimittelforsch.*, *37*, 1197-1201.

Dzondo-Gadet, M., Mayap-Nzietchueng, R., Hess, K., Nabet, P., Belleville, F., Dousset, B. (2002). Action of boron at the molecular level: effects on transcription and translation in an acellular system. *Biol. Trace Elem. Res.*, *85*, 23-33.

Eren, M., Uyanik, F. (2007). Influence of dietary boron supplementation on some serum metabolites and egg-yolk cholesterol in laying hens. *Acta Vet. Hung.*, *55*, 29-39.

Fail, P.A., Chapin, R.E., Price, C.J., Heindel, J.J. (1998). General reproductive, developmental and endocrine toxicity of boronated compounds. *Reprod. Toxicol.*, *12*, 1-18.

Fail, P.A., George, J.D., Seely, J.C., Grizzle, T.B., Heindel, J.J. (1991). Reproductive toxicity of boric acid in Swiss (CD-1) mice: assessment using the continuous breeding protocol. *Fundam. Appl. Toxicol.*, *17*, 225-239.

Fort, D.J., Stover, E.L., Bantle, J.A., Dumont, J.N., Finch, R.A. (2001). Evaluation of a reproductive toxicity assay using *Xenopus laevis*: boric acid, cadmium and ethylene glycol monomethyl ether. *J. Appl. Toxicol.*, *21*, 41-52.

Fukuda, R., Hiroda, M., Mori, I., Chatani, F., Morishima, H., Mayahare, H. (2000). Collaborative work to evaluate toxicity on male reproductive organs by repeated dose studies in rats 24. Testicular toxicity of boric acid after 2- and 4- week administration periods. *J. Toxicol. Sci.*, *25*, 233-239.

Gallardo-Williams, M.T., Maronpot, R.R., Wine, R.N., Brunssen, S.H., Chapin, R.E. (2003). Inhibition of the enzymatic activity of prostate-specific antigen by boric acid and 3-nitrophenyl boronic acid. *Prostate*, *54*, 44-49.

Gosselin, R.E., Hodge, N.C., Smith, R.P., Eds. (1976). *Clinical Toxicology of Commercial Products*, 4th ed., Baltimore, MD: Williams & Wilkins, pp. 63-66.

Hamilton, R.A., Wolf, B.C. (2007). Accidental boric acid poisoning following the ingestion of household pesticide. *J. Forensic Sci.*, *52*, 706-708.

Harris, M.W., Chapin, R.E., Lockhart, A.C., Jokinen, M.D. (1992). Assessment of a short-term reproductive and developmental toxicity screen. *Fundam. Appl. Toxicol.*, *19*, 181-196.

Harrouk, W.A., Wheeler, K.E., Kimmel, G.L., Hogan, K.A., Kimmel, C.A. (2005). Effects of hyperthermia and boric acid on skeletal development in rat embryos. *Birth Defects Res. B. Dev. Reprod. Toxicol.*, *74*, 268-276.

Heindel, J.J., Price, C.J., Schwetz, B.A. (1994). The developmental toxicity of boric acid in mice, rats and rabbits. *Environ. Health Perspect.*, *102(Suppl.7)*, 107-112.

Heindel, J.J., Price, C.J., Field, E.A., Marr, M.C., Myers, C.B., Morrissey, R.E., Schwetz, B.A. (1992). Developmental toxicity of boric acid in mice and rats. *Fundam. Appl. Toxicol.*, *18*, 266-277.

Hubbard, S.A. (1998). Comparative toxicology of borates. *Biol. Trace Elem. Res.*, *66*, 343-357.

Hyrsl, P., Büyükgüzel, E., Büyükgüzel, K. (2007). The effects of boric acid-induced oxidative stress on antioxidant enzymes and survivorship in *Galleria mellonella*. *Arch. Insect Biochem. Physiol.*, *66*, 23-31.

Hyrsl, P., Büyükgüzel, E., Büyükgüzel, K. (2008). Boric acid-induced effects on protein profiles of *Galleria mellonella* hemolymph and fat body. *Acta Biol. Hung.*, *59*, 281-288.

Ishii, Y., Fujizuka, N., Takahashi, T., Shimizu, K., Tuchida, A., Yano, S., Naruse, T., Chishiro, T. (1993). A fatal case of acute boric acid poisoning. *J. Toxicol. Clin. Toxicol.*, *31*, 345-352.

Jansen, J.A., Schon, J.S., Aggerbeck, B. (1984a). Gastro-intestinal absorption and in vitro release of boric acid from water-emulsifying ointments. *Food Chem. Toxicol.*, *22*, 44-53.

Juszkiewicz, A., Zaborska, W., Sepioł, J., Góra, M., Zaborska, A. (2003). Inactivation of jack bean urease by allicin. *J. Enzyme Inhib. Med. Chem.*, *18*, 419-424.

Kelani, K.M. (2004). Selective potentiometric determination of zolpidem hemitartrate in tablets and biological fluids by using polymeric membrane electrodes. *J. AOAC Int.*, *87*, 1309-1318.

Kiesche-Nesselrodt, A., Hooser, A. (1990). Toxicology of selected pesticides, drugs, and chemicals. Boric acid. *Vet. Clin. North Am. Small Anim. Pract.*, *20*, 369-373.

Kikuchi, T., Suzuki, M., Kusai, A., Iseki, K., Sasaki, H. (2005a). Synergistic effect of EDTA and boric acid on corneal penetration of CS-088. *Int. J. Pharm.*, *290*, 83-89.

Kikuchi, T., Suzuki, M., Kusai, A., Iseki, K., Sasaki, H., Nakashima, K. (2005b). Mechanism of permeability-enhancing effect of EDTA and boric acid on the corneal penetration of 4-[1-hydroxy-1-methylethyl]-2-propyl-1-[4-[2-[tetrazole-5-yl] phenyl] phenyl] methylimidazole-5-carboxylic acid monohydrate (CS-088). *Int. J. Pharm.*, *299*, 107-114.

Kim, D.H., Hee, S.Q., Norris, A.J., Faull, K.F., Eckhert, C.D. (2006). Boric acid inhibits adenosine diphosphate-ribosyl cyclase non-competitively. *J. Chromatogr. A.*, *1115*, 246-252.

Kot, M., Zaborska, W. (2003). Irreversible inhibition of jack bean urease by pyrocatechol. *J. Enzyme Inhib. Med. Chem.*, *18*, 413-417.

Ku, W.W., Chapin, R.E. (1994). Mechanism of the testicular toxicity of boric acid in rats: in vivo and in vitro studies. *Environ. Health Perspect., 102 (Suppl.7)*, 99-105.

Ku, W.W., Shih, L.M., Chapin, R.E. (1993a). The effects of boric acid on testicular cells in culture. *Reprod. Toxicol., 7*, 321-331.

Ku, W.W., Chapin, R.E., Wine, R.N., Gladen, B.C. (1993b). Testicular toxicity of boric acid: relationship of dose to lesion development and recovery in the F344 rat. *Reprod. Toxicol., 7*, 305-319.

Kudo, S., Tanase, H., Yamasaki, M., Nakao, M., Miyata, Y., Tsuru, K., Imai, S. (2000). Collaborative work to evaluate toxicity on male reproductive organs by repeated dose studies in rats: 23. A comparative 2- and 4- week repeated oral dose testicular toxicity study of boric acid in rats. *J. Toxicol. Sci., 25*, 223-232.

Letondor, C., Humbert, N., Ward, T.R. (2005). Artificial metalloenzymes based on biotin-avidin technology for the enantioselective reduction of ketones by transfer hydrogenation. *Proc. Natl. Acad. Sci., 102*, 4683-4687.

Li, J., Jiang, Y., Sun, T., Ren, S. (2008). Fast and simple method for assay of ciclopirox olamine by micellar electrokinetic capillary chromatography. *J. Pharm. Biomed. Anal., 47*, 929-933.

Linden, C.H., Hall, A.H., Kulig, K.W., Rumack, B.H. (1986). Acute ingestions of boric acid. *J. Toxicol. Clin. Toxicol., 24*, 269-279.

Linder, R.E., Strader, L.F., Rehnberg, G.L. (1990). Effect of acute exposure to boric acid on the male reproductive system of the rat. *J. Toxicol. Environ. Health, 31*, 133-146.

Litovitz, T.L., Klein-Schwartz, W., Oderda, G.M., Schmitz, B.F. (1988). Clinical manifestations of toxicity in a series of 784 boric acid ingestions. *Am. J. Emerg. Med., 6*, 209-213.

Liu, J., Lou, Y.J. (2004). Determination of icariin and metabolites in rat serum by capillary zone electrophoresis: rat pharmacokinetic studies after administration of icariin. *J. Pharm. Biomed. Anal., 36*, 365-370.

Locatelli, C., Minoia, C., Tonini, M., Manzo, L. (1987). Human toxicology of boron with special reference to boric acid poisoning. *G. Ital. Med. Lev., 9*, 141-146.

Macholan, L., Skladal, P., Bohackova, I., Krejci, J. (1992). Amperometric glucose biosensor with extended concentration range utilizing complexation effect of borate. *Biosens. Bioelectron.*, 7, 593-598.

Maeda, E., Kataoka, M., Hino, M., Kajimoto, K., Kaji, N., Tokeshi, M., Kido, J., Shinohara, Y., Baba, Y. (2007). Determination of human blood glucose levels using microchip electrophoresis. *Electrophoresis*, 28, 2927-2933.

Matsuda, K., Okamoto, M., Ashida, M., Ishimaru, T., Horiuti, I., Suzuki, K., Yamamoto, S. (2004). Toxicological analysis over the past five years at a single institution, *Rinsho Byori (Japan)*, 52, 819-823.

Mendham, J., Denney, R.C., Barnes, J.D., Thomas, M.J.K. (2000). *Vogel's Textbook of Quantitative Chemical Analysis*, 6th ed., Delhi: Dorling Kindersley, pp. 384-385.

Moffat, A.C., Osselton, M.D., Widdop, B., Eds. (2004). *Clarke's Analysis of Drugs and Poisons,* 3rd ed., London: Pharmaceutical Press, p. 269.

Moore, J.A. (1997). An assessment of boric acid and borax using the IEHR evaluative process for assessing human developmental and reproductive toxicity of agents. Expert Scientific Committee. *Reprod. Toxicol., 11*, 123-160.

Morten, J.A., Delves, H.T. (1999). Measurement of total boron and ^{10}B concentration and the detection and measurement of elevated ^{10}B levels in biological samples by inductively coupled plasma-mass spectrometry using the determination of ^{10}B: ^{11}B ratios. *J. Anal. Atomic Spectrom., 14*, 1545-1556.

Murray, F.J. (1995). A human health risk assessment of boron (boric acid and borax) in drinking water. *Regul. Toxicol. Pharmacol., 22*, 221-230.

Narotsky, M.G., Schmid, J.E., Andrews, J.E., Kavlock, R.J. (1998). Effects of boric acid on axial skeletal development in rats. *Biol. Trace Elem. Res., 66*, 373-394.

Odunola, O.A. (1997). Individual and combined genotoxic response of boric acid and aflatoxin B1 in *Escherichia coli* PQ 37. *East Afr. Med. J., 74*, 499-502.

Pacenti, M., Dugheri, S., Villanelli, F., Bartolucci, G., Calamai, L., Boccalon, P., Arcangeli, G., Vecchione, F., Alessi, P., Kikic, I., Cupelli, V. (2008). Determination of organic acids in urine by solid-phase microextraction and gas chromatography-ion trap tandem mass spectrometry previous 'in sample' derivatization with trimethyloxonium tetrafluoroborate. *Biomed. Chromatogr.*, 22,1155-1163.

Pahl, M.V., Culver, B.D., Vaziri, N.D. (2005). Boron and the kidney. *J. Ren. Nutr., 15*, 362-370.

Pahl, M.V., Culver, B.D., Strong, D.L., Murray, F.J., Vaziri, N.D. (2001). The effect of pregnancy on renal clearance of boron in humans: a study based on normal dietary intake of boron. *Toxicol. Sci., 60*, 252-256.

Pascali, J.P., Liotta, E., Gottardo, R., Bortolotti, F., Tagliaro, F. (2009). Rapid optimized separation of bromide in serum samples with capillary zone electrophoresis by using glycerol as additive to the background electrolyte. *J. Chromatogr. A., 1216*, 3349-3352.

Peters, A.K., Steemans, M., Hansen, E., Mesens, N., Verheyen, G.R., Vanparys, P. (2008a). Evaluation of the embryotoxic potency of compounds in a newly revised high throughput embryonic stem cell test. *Toxicol. Sci., 105*, 342-350.

Peters, A.K., Wouwer, G.V., Weyn, B., Verheyen, G.R., Vanparys, P., Gompel, J.V. (2008b). Automated analysis of contractility in the embryonic stem cell test, a novel approach to assess embryotoxicity. *Toxicol. In Vitro, 22*, 1984-1956.

Pongsavee, M. (2009). Genotoxic effects of borax on cultured lymphocytes. *Southeast Asian J. Trop. Med. Public Health, 40*, 411-418.

Price, C.J., Strong, P.L., Murray, F.J., Goldberg, M.M. (1997). Blood boron concentrations in pregnant rats fed boric acid throughout gestation. *Reprod. Toxicol., 11*, 833-842.

Price, C.J., Strong, P.L., Murray, F.J., Goldberg, M.M. (1998). Developmental effects of boric acid in rats related to maternal blood boron concentrations. *Biol. Trace Elem. Res., 66*, 359-372.

Price, C.J., Strong, P.L., Marr, M.C., Mayers, C.B., Murray, F.J. (1996b). Developmental toxicity NOAEL and postnatal recovery in rats fed boric acid during gestation. *Fundam. Appl. Toxicol., 32*, 179-193.

Price, C.J., Marr, M.C., Myers, C.B., Seely, J.C., Heindel, J.J., Schwetz, B.A. (1996a). The developmental toxicity of boric acid in rabbits. *Fundam. Appl. Toxicol., 34*, 176-187.

Rauniyar, V., Hall, D.G. (2009). Rationally improved chiral Brønsted acid for catalytic enantioselective allylboration of aldehydes with an expanded reagent scope. *J. Org. Chem., 74*, 4236-4241.

Reddy, K.R., Kayastha, A.M. (2006). Boric acid and boronic acids inhibition of pigeonpea urease. *J. Enzyme Inhib. Med. Chem., 21*, 467-470.

Reilly, W.J. Jr. (2006). Pharmaceutical necessities, In: R. Hendrickson, Ed., *Remington: The Science and Practice of Pharmacy*, 21st ed., Philadelphia, PA: Lippincott Williams & Wilkins, pp. 1083-1084, 1089.

Restuccio, A., Mortensen, M.E., Kelly, M.T. (1992). Fatal ingestion of boric acid in an adult. *Am. J. Emerg. Med., 10*, 545-547.

Richhold, M. (1998). Boron exposure from consumer products. *Biol. Trace Elem. Res., 66*, 121-129.

Robbins, W.A., Xun, L., Jia, J., Kennedy, N., Elashoff, DA, Ping, L. (2009). Chronic boron exposure and human semen parameters. *Reprod. Toxicol., 29*, 184-190.

Robinson-Fuentes, V.A., Jaime-Sánchez, J.L., García-Aguilar, L., Gómez-Peralta, M., Vázquez-Garcidueñas, M.S., Vázquez-Marrufo, G. (2008). Determination of alpha- and beta-amanitin in clinical urine samples by Capillary Zone Electrophoresis. *J. Pharm. Biomed. Anal., 47*, 913-917.

Rowe, R.I., Bouzan, C., Nabili, S., Eckert, C.D. (1998). The response of trout and zebrafish embryos to low and high boron concentrations is U-shaped. *Biol. Trace Elem. Res., 66*, 261-270.

Saxena, R., Verma, R.M. (1983). Iodometric microdetermination of boric acid and borax separately or in a mixture. *Talanta, 30*, 365-367.

Seigel, E., Wilson, S. (1986). Boric acid toxicity. *Pediatr. Clin. North Am., 33*, 363-367.

Shomron, N., Ast, G. (2003). Boric acid reversibly inhibits the second step of pre-mRNA splicing. *FEBS Lett., 552*, 219-224.

Soine, T.O., Wilson, C.O. (1967). *Rogers Inorganic Pharmaceutical Chemistry*, 8th ed., Philadelphia, PA: Lea & Febiger, pp. 125-126.

Stuttgen, G., Siebel, T., Aggerbeck, B. (1982). Absorption of boric acid through skin depending on the type of vehicle. *Arch. Dermatol. Res., 272*, 21-29.

Suhara, Y., Takayama, H., Nakane, S., Miyashita, T., Waku, K., Sugiura, T. (2000). Synthesis and biological activities of 2-arachidonoylglycerol, an endogenous cannabinoid receptor ligand, and its metabolically stable ether-linked analogues. *Chem. Pharm. Bull. (Tokyo), 48*, 903-907.

Tangermann, R.H., Etzel, R.A., Mortimer, L., Penner, G.D., Paschal, D.C. (1992). An outbreak of a food related illness resembling boric acid poisoning. *Arch. Environ. Contam. Toxicol., 23*, 142-144.

Teshima, D., Taniyama, T., Oishi, R. (2001). Usefulness of forced diuresis for acute boric acid poisoning in an adult. *J. Clin. Pharm. Ther., 26*, 387-390.

Torregroza, I., Evans, T., Das, B.C. (2009). A forward chemical screen using zebrafish embryos with novel 2-substituted 2H-chromene derivatives. *Chem. Biol. Drug Des., 73*, 339-345.

Treinen, K.A., Chapin, R.E. (1991). Development of testicular lesions in F 344 rats after treatment with boric acid. *Toxicol. Appl. Pharmacol., 107*, 325-335.

Turkez, H. (2008). Effects of boric acid and borax on titanium dioxide genotoxicity. *J. Appl. Toxicol., 28*, 658-664.

Turkez, H., Tatar, A., Hacimuftuoglu, A., Ozdemir, E. (2010). Boric acid as a protector against paclitaxel genotoxicity. *Acta Biochim. Pol., 57*, 95-97.

Turkez, H., Geyikoglu, F., Tatar, A., Keles, S., Ozkan, A. (2007). Effects of some boron compounds on peripheral human blood. *Z. Naturforsch. C., 62*, 889-896.

Turkoglu, S. (2007). Genotoxicity of five food preservatives tested on root tips of Allium cepa L. *Mutat. Res., 626*, 4-14.

Tyl, R.W., Chernoff, N., Rogers, J.M. (2007). Altered axial skeletal development. *Birth Defects Res. B Dev. Reprod. Toxicol., 80*, 451-472.

Vaziri, N.D., Oveisi, F., Culver, B.D., Pahl, M.V., Anderson, M.E., Strong, P.L., Murray, F.J. (2001). The effect of pregnancy on renal clearance of boron in rats given boric acid orally. *Toxicol. Sci., 60*, 257-263.

Vieira, A.S., Fiorante, P.F., Hough, T.L., Ferreira, F.P., Lüdtke, D.S., Stefani, H.A. (2008). Nucleophilic addition of potassium alkynyltrifluoroborates to D-glucal mediated by BF3 x OEt2: highly stereoselective synthesis of alpha-C-glycosides. *Org. Lett., 10*, 5215-5218.

Wang, Y., Zhao, Y., Chen, X. (2008). Experimental study on the estrogen-like effect of boric acid. *Biol. Trace Elem. Res., 121*, 160-170.

Werfel, S., Boeck, K., Abeck, D., Ring, J. (1998). Special characteristics of topical treatment in childhood. *Hautarzt, 49*, 170-175.

Wery, N., Narotsky, M.G., Pacico, N., Kavlock, R.J., Picard, J.J., Gofflot, F. (2003). Defects in cervical vertebrae in boric acid-exposed rat embryos are associated with anterior shifts of hox gene expression domains. *Birth Defects Res. A, Clin. Mol. Teratol., 67*, 59-67.

Wester, R.C., Hartway, T., Maibacch, H.I., Schell, M.J., Northington, D.J., Culver, B.D., Strong, P.L. (1998a). In vitro percutaneous absorption of boron as boric acid, borax, and disodium octaborate tetrahydrate in human skin: a summary. *Biol. Trace Elem. Res., 66*, 111-120.

Wester, R.C., Hui, X., Maibach, H.I., Bell, K., Schell, M.J., Northington, D.J., Strong, P., Culver, B.D. (1998b). In vivo percutaneous absorption of boric acid, borax, and disodium octaborate tetrahydrate in humans: a summary. *Biol. Trace Elem. Res., 66,* 101-109.

Wester, R.C., Hui, X., Hartway, T., Maibach, H.I., Bell, K., Schell, M.J., Northington, D.J., Strong, E., Culver, B.D. (1998c). In vitro percutaneous absorption of boric acid, borax, and disodium octaborate tetrahydrate in humans compared to in vitro absorption in human skin from infinite and finite doses. *Toxicol. Sci., 45*, 42-51.

Wise, L.D., Winkelmann, C.T. (2009). Micro-computed tomography and alizarin red evaluations of boric acid-induced fetal skeletal changes in Sprague-Dawley rats. *Birth Defects Res. B Dev. Reprod. Toxicol.*, *86*, 214-219.

Woods, W.G. (1994). An introduction to boron: history, sources, uses and chemistry. *Environ. Health Perspect.*, *102*, 5-11.

Yoshida, M., Watabiki, T., Ishida, N. (1989). Spectrophotometric determination of boric acid by the curcumin method. *Nihon Hoigaku Zasshi*, *43*, 490-496.

Yoshizaki, H., Izumi, Y., Hirayama, C., Fujimoto, A., Kandori, H., Sugitani, T., Ooshima, Y. (1999). Availability of sperm examination for male reproductive toxicities in rats treated with boric acid. *J. Toxicol. Sci.*, *24*, 199-208.

Zhao, S., Wang, J., Ye, F., Liu, Y.M. (2008). Determination of uric acid in human urine and serum by capillary electrophoresis with chemiluminescence detection. *Anal. Biochem.*, *378*, 127-131.

Zinellu, A., Carru, C., Sotgia, S., Deiana, L. (2004). Optimization of ascorbic and uric acid separation in human plasma by free zone capillary electrophoresis ultraviolet detection. *Anal. Biochem.*, *330*, 298-305.

Zinellu, A., Pinna, A., Zinellu, E., Sotgia, S., Deiana, L., Carru, C. (2008). High-throughput capillary electrophoresis method for plasma cysteinylglycine measurement: evidences for a clinical application. *Amino Acids*, *34*, 69-74.

Chapter 5

APPLICATIONS OF BORATES

Boric acid and borates have extensive industrial and consumer applications. Reviews on a wide range of industrial applications (Rosenfelder, 1978; Woods, 1994; Garrett, 1998; O'Neil, 2001; Flores, 2004; Chung, 2010), chemical and biological applications (Ali *et al.*, 2005), pharmaceutical and medical applications (Garrett, 1998; Reilly, 2006; Ahmad *et al.*, 2010a,b) and as wood preservatives (Unger *et al.*, 2001; Kartal, 2010) and fire retardants (Garrett, 1998; Wang *et al.*, 2004) have been published. Some of the important applications of these compounds described by the above authors are presented in the following sections.

5.1. GENERAL APPLICATIONS

5.1.1. Boric Acid

Boric acid is used for weatherproofing wood and fireproofing fabrics; manufacturing cements, crockery, porcelain, enamels, glass, borates, leather, carpets, hats, soaps, artificial gems; in nickeling baths; cosmetics; printing and dyeing, painting; photography; for impregnating wicks; electric condensers and hardening steel.

5.1.2. Borates

A. Sodium Borate (Borax)

It is used for soldering metals; manufacturing glazes and enamels; tanning; in cleaning compounds; artificially aging wood; fireproofing fabrics and wood and curing skins. In water it forms an alkaline solution and precipitates aluminum salts as aluminum hydroxide, iron salts as basic borates and zinc sulfate as zinc borate. Alkaloids are precipitated from the solutions of their salts.

B. Sodium Perborate

Sodium perborate is used for bleaching straw and other fibers, ivory, sponges, bristles, waxes, textiles; in laundering, dentifrices and soaps.

C. Calcium Borate

It is used as flux in heavy-metal metallurgy; in manufacturing forsterite porcelain insulators; in glycol antifreezes; in fire-retardant paints.

D. Magnesium Borate

It is an antiseptic and fungicide.

E. Disodium Octaborate

It is a wood preservative.

F. Copper Borate

It is a wood preservative.

G. Zinc Borate

It is used as a fire retardant for wood.

H. Phenylmercuric Borate

It is an antimicrobial preservative.

I. Potassium / Sodium Borohydride

These are used as reducing agents in chemical reactions such as the synthesis of corticosteroids.

J. Boron Trifluoride

It is used as Lewis acid in condensation reactions.

5.2. MISCELLANEOUS APPLICATIONS

5.2.1. Preservatives

Boron-based wood preservatives and their uses have been reviewed by Kartal (2010). Various species of bacteria, fungi and insects are known to be affected by high concentrations of borate compounds. Vitamins and coenzymes present in the biological system can react with boron to form stable complexes and thus affect the metabolic processes (Lloyd et al., 1990; Lloyd and Dickinson, 1992). The fungicidal action of borates involves the interaction of oxidized coenzymes NAD^+, NMN^+ and $NADP^+$ with borate ions (Lloyd, 1998). Borates interfere with the digestive processes of insects and are thus toxic to them (Maistrello et al., 2003). Boric acid, sodium borate, disodium octaborate, copper borate and zinc borate are widely used in household and waterborne wood preservation against fungi and insects (Unger et al., 2001; Kartal, 2010).

Boric acid and borates are used against decay and rot fungi (*Postia placenta* and *Coniophora puteana*) (Williams and Amburgey, 1987; Temiz et al., 2008). [11]B nuclear magnetic resonance imaging and spectroscopy have been used to characterize the nature and distribution of boron compounds after preservative treatment of radiate pine wood with trimethylborate. Trimethylborate undergoes rapid hydrolysis to form boric acid in pine wood (Meder et al., 1999). Strong fungicide and insecticide effects have been observed on treating wood with composite films of boric acid (Bottcher et al., 1999). Inorganic borates offer good protection to timber in most non-ground contact applications (Obanda et al., 2008). Fungal degradation of paper sheets may be used for screening different wood preservatives including boric acid on paper instead of solid wood (Raberg and Hafra, 2009). Urine samples collected at home and in the hospital are preserved with boric acid for storage before processing (Gillespie et al., 1999; Thierauf et al., 2008).

Borates have been used in mummification processes in Pharaonic Egypt more than 4000 years ago. Salt samples of borates used as embalming material for mummies contained 2-4 µM/g. Smaller borate concentrations (1.2 µM/g) have been found in ancient bone samples (Kaup et al., 2003).

5.2.2. Fire Retardants

Boric acid and borates have been used as fire retardant in wood preservation industry (Baysal *et al.*, 2007). The application of borates in preparing inexpensive cellulose insulation material by treating cellulose with a borate solution to give a fire resistant material is described (Garrett, 1998). The reaction of boric acid with hydroxyl groups of cellulose results in the formation of a thin and stable film (Rosenfelder, 1978). Zinc borate has been found to possess greater flame retardancy than other borates. It promotes char formation with inhibition of the release of combustible material. Zinc borate and aluminum trihydrate form a synergestic mixture to reduce smoke (Lyday, 1985).

Borax and boric acid can be incorporated in to particle board chips to produce a fire retardant hard board (Kartal, 2010). Borax tends to reduce flame spread and boric acid suppresses glowing and, therefore, both are generally used together for fireproofing (Wang *et al.*, 2004).

5.2.3. Insecticides

Insecticidal use of borates is widespread because of their low mammalian toxicity and negligible insect resistance. Boric acid and borax have been used as insecticides in the control of American cockroaches (Lizzio, 1986), German cockroaches (*Blattella germanica*) (Strong *et al.*, 1993; Cochran, 1995; Miller and Kochler, 2000; Habes *et al.*, 2001; Zurek *et al.*, 2003; Gore and Schal, 2004; Gore *et al.*, 2004; Appel *et al.*, 2004; Zhang *et al.*, 2005; Kilani-Morakchi *et al.*, 2006; Wang and Bennett, 2009), adult mosquitoes (Diptera: Culicidae) (Xue and Barnard, 2003; Ali *et al.*, 2006; Xue *et al.*, 2006, 2008), cat fleas (Siphonaptera: Pulicidae) (Klotz *et al.*, 1994), house flies (Diptera: Muscidae) (Hogsette *et al.*, 2002), fruit flies (Diptera: Tephritidae) (Heath *et al.*, 2009) and Argentine ants (Hymenoptera: Formicidae) (Hooper-Bui and Rust, 2000; Klotz *et al.*, 2000; Ulloa-Chacon and Jaramillo, 2003; Rust *et al.*, 2004; Stanley and Robinson, 2007; Daane *et al.*, 2008; Chong and Lee, 2009; Brightwell and Silverman, 2009). In the larvae of the wax moth (*Galleria mellonella*) the boric acid toxicity is related, in part, to oxidative stress management (Hyrsl *et al.*, 2007). The boric acid induced effects on protein profiles of the wax moth indicated a marked quantitative change in the 45 kDa protein fraction of the hemolymph in the VIIth instar larvae reared on 2500 ppm dietary boric acid (Hyrsl *et al.*, 2008). A 0.3% concentration of sodium tetraborate decreased fat body lysozyme of the larvae from 0.12 ± 0.013 to 0.006 ± 0.003 mg / mL in VIIth instar (Durmus and Büyükgüzel, 2008).

Morphological alterations induced by boric acid in the midgut of honeybee (*Apis mellifera* L.) larvae have been observed (da Silva Cruz *et al.*, 2010). The effects of various boron compounds including boric acid, borax, zinc borate, or sodium perborate tetrahydrate on the termite resistance have shown that highest mortalities occur with either boric acid or borax (Usta *et al.*, 2009).

5.2.4. Pharmaceuticals

Boric acid, sodium borate and sodium perborate are the pharmaceutical necessities cited in the British Pharmacopoeia (2009) and the United States Pharmacopeia-National Formulary (2007). These compounds are mild antiseptics and inhibit Gram negative bacteria. Boric acid is an officially recognized buffer in the pharmacopoeias. Aqueous solutions of boric acid are used as an eye wash, mouth wash and for irrigation of the bladder. It is also used as a dusting powder, diluted with some inert material. Sodium borate is used as an alkalizing agent and as a buffer for alkaline solutions (Reilly, 2006). Sodium perborate has been used as a bleaching agent for discolored teeth (Baumler *et al.*, 2006; Cavalli *et al.*, 2009; Valera *et al.*, 2009). Boric acid has also been used as an antibacterial agent in wound treatment, antifungal agent for infections caused by *Candida* species, chemopreventive agent for human cancer and for the treatment of ear infections (Ahmad *et al.*, 2010a,b). Boric acid is used in the pharmaceutical industry in various preparations and as a raw material to obtain other boron products of higher added value such as sodium perborate, ammonium pentaborate, etc. (Flores and Valdez, 2010).

5.2.5. Cosmetics

The Greek physician, Galen, in 200 A.D. developed a cosmetic preparation containing rose water, bees wax and olive oil (DeNavarre, 1975). Several other oil combinations and creams have been used by early Egyptians to protect against hot sun and dry winds (Chaudhri and Jain, 2009). The basic formula of Galen has been used for centuries. In the late 1800s borax was a valuable addition to the basic formula to prepare a simple cold cream formulation (Wilmott *et al.*, 2005). A number of cosmetic preparations containing boric acid / borax have been reported (Vimaladevi, 2005). Sodium and potassium borates are used in face creams, lotions, dusting powders, ointments, hair preparations and as emulsifiers in cosmetic formulations (Kistler and Helvaci, 1994). Intoxication due to the use of cosmetics containing boric acid and borax has been observed (Larcan *et al.*, 1975). Lethal complications, allergic contact eczemas and toxic irritations of the skin are caused by the use of cosmetics formulated with boric acid (Haustein and Barth, 1983).

5.2.6. Borosilicate Glasses

Borosilicate glass (Pyrex) is used as a material of choice for containers to dispense a variety of pharmaceutical preparations including both parenteral and non-parenteral formulations. It is composed principally of silicon dioxide with varying concentration of boron and other oxides such as sodium, potassium, calcium, magnesium, aluminum, iron, etc. In this glass the main network is formed by silicon dioxide and then the spaces in the network are filled by boron oxide and other oxides. According to USP-NF, the different types of glasses required for various pharmaceutical preparations, Type I consist of borosilicate glass. It is the most unreactive of all the glasses and is used for water for injection and those products requiring terminal sterilization. (Avis, 1986; Abendroth and Clark, 1992; Amberosio, 2002; United States Pharmacopeia-National Formulary, 2007; Flores and Valdez, 2010).

Borosilicate glass possesses increased thermal shock resistance because anhydrous boric acid (B_2O_3) lowers its expansion coefficient. Boric oxide in fiber glass produces the desirable drawing qualities and increases the mechanical strength (Woods, 1994).

REFERENCES

Abendroth, R.P., Clark, R.N. (1992). Glass containers for parenterals, In: K.E. Avis, H.A. Lieberman, L. Lachman, Eds., *Pharmaceutical Dosage Forms: Parenteral Medications*, Vol. 1, New York, NY: Marcel Dekker, Inc., Chap. 9.

Ahmad, I., Ahmed, S., Sheraz, M.A., Vaid, F.H.M. (2010a). Borate: toxicity, effect on drug stability and analytical applications, In: M.P. Chung, Ed.,

Handbook on Borates: Chemistry, Production and Applications, New York, NY: Nova Science Publishers Inc., Chap. 2.

Ahmad, I., Ahmed, S., Sheraz, M.A., Iqbal, K., Vaid, F.H.M. (2010b). Pharmacological aspects of borates. *Int. J. Med. Biol. Front., 16,* 977-1004.

Ali, H.A., Dembistsky, V.M., Srebnisk, M. (2005). *Studies in Inorganic Chemistry. 22 Contemporary Aspects of Boron: Chemical and Biological Applications*, Amsterdam: Elsevier.

Ali, A., Xue, R.D., Barnard, D.R. (2006). Effects of sublethal exposure to boric acid sugar bait on adult survival, host-seeking, blood feeding behavior, and reproduction of Stegomyia albopicta. *J. Am. Mosq. Control Assoc., 22,* 464-468.

Amberosio, T.J. (2002). Packaging of pharmaceutical dosage forms, In: G.S. Banker, C.T. Rhode, Eds., *Modern Pharmaceutics*, 4th ed., New York, NY: Marcel Dekker, Inc., Chap 17.

Appel, A.G., Gehret, M.J., Tanley, M.J. (2004). Effects of moisture on the toxicity of inorganic and organic insecticidal dust formulations to German cockroaches (Blattodea: Blattellidae). *J. Econ. Entomol., 97,* 1009-1016.

Avis, K.E. (1986). Sterile products, In: L. Lachman, H.A. Lieberman, J.L. Kanig, Eds., *The Theory and Practice of Industrial Pharmacy*, Philadelphia: Lea & Febiger, Chap 22.

Bäumler, M.A., Schug, J., Schmidlin, P., Imfeld, T. (2006). In vitro tests of internal tooth whitening agents on colored solutions do not replace tests on teeth. *Schweiz. Monatsschr. Zahnmed., 116,* 1000-1005.

Baysal, E., Altinok, M., Colak, M., Ozaki, S.K., Toker, H. (2007). Fire resistance of Douglas fir (*Pseudotsuga menzieesi*) treated with borates and natural extractives. *Bioresour. Technol., 98,* 1101-1105.

Böttcher, H., Jagota, C., Trepte, J., Kallies, K.H., Haufe, H. (1999). Sol-gel composite films with controlled release of biocides. *J. Control Release, 60,* 57-65.

Brightwell, R.J., Silverman, J. (2009). Effects of honeydew-producing hemipteran denial on local argentine ant distribution and boric acid bait performance. *J. Econ. Entomol., 102,* 1170-1174.

British Pharmacopoeia (2009). London: Her Majesty's Stationary Office, Electronic Version.

Cavalli, V., Shinohara, M.S., Ambrose, W., Malafaia, F.M., Pereira, P.N., Giannini, M. (2009). Influence of intracoronal bleaching agents on the ultimate strength and ultrastructure morphology of dentine. *Int. Endod. J., 42,* 568-575.

Chaudhri, S.K., Jain, N.K. (2009). History of cosmetics. *Asian J. Pharm.*, *3*, 164-167.

Chong, K.F., Lee, C.Y. (2009). Evaluation of liquid baits against field populations of the longlegged ant (Hymenoptera: Formicidae). *J. Econ. Entomol.*, *102*, 1586-1590.

Chung, M.P., Ed. (2010). *Handbook on Borates: Chemistry, Production and Applications*, New York, NY: Nova Science Publishers Inc.

Cochran, D.G. (1995). Toxic effects of boric acid on the German cockroach. *Experientia, 51*, 561-563.

da Silva Cruz, A., da Silva-Zacarin, E.C., Bueno, O.C., Malaspina, O. (2010). Morphological alterations induced by boric acid and fipronil in the midgut of worker honeybee (*Apis mellifera* L.) larvae: Morphological alterations in the midgut of *A. mellifera*. *Cell Biol. Toxicol.*, *26*, 165-176.

Daane, K.M., Cooper, M.L., Sime, K.R., Nelson, E.H., Battany, M.C., Rust, M.K. (2008). Testing baits to control Argentine ants (Hymenoptera: Formicidae) in vineyards. *J. Econ. Entomol.*, *101*, 699-709.

DeNavarre, M.G., Ed. (1975). *The Chemistry and Manufacture of Cosmetics*, Orlando, FL: Continental Press, p.237.

Durmus, Y., Büyükgüzel, K. (2008). Biological and immune response of *Galleria mellonella* (Lepidoptera: Pyralidae) to sodium tetraborate. *J. Econ. Entomol.*, *101*, 777-783.

Flores, H.R. (2004). *El Beneficio de los Boratos, Historia, Minerales, Yacimientos, Usos, Tratamiento, Refinacion, Propiedades, Contaminacion, Analisis Quimico*. Crisol Ediciones, Salta, Argentina.

Flores, H., Valdez, S. (2010). Purification of boric acid by washing, In: M.P. Chung, Ed., *Handbook on Borates: Chemistry, Production and Applications*, New York, NY: Nova Science Publishers Inc., Chap. 14.

Garrett, D.E. (1998). *Borates Handbook of Deposits, Processing, Properties and Uses*, San Diago, CA:Academic Press, Chap. 9.

Gillespie, T., Fewster, J., Masterton, R.G. (1999). The effect of specimen processing delay on borate urine preservation. *J.Clin. Pathol.*, *52*, 95-98.

Gore, J.C., Schal, C. (2004). Laboratory evaluation of boric acid-sugar solutions as baits for management of German cockroach infestations. *J. Econ. Entomol., 97,* 581-587.

Gore, J.C., Zurek, L., Santangelo, R., Stringham, S.M., Watson, D.W., Schal, C. (2004). Water solutions of boric acid and sugar for management of German cockroach populations in live stock production systems. *J. Econ. Entomol., 97,* 715-720.

Habes, D., Kilani-Morakchi, S., Aribi, N., Farine, J.P., Soltani, N. (2001). Toxicity of boric acid to *Blattella germanica* (Dictyoptera: Blattellidae) and analysis of residues in several organs. *Meded Rijksuniv Gent Fak Landbouwkd Toegep Biol. Wet., 66,* 525-534.

Haustein, U.F., Barth, J. (1983). Side effects of dermatologic local therapy. *Z. Gesamte Inn. Med., 38,* 663-668.

Heath, R.R., Vazquez, A., Schnell, E.Q., Villareal, J., Kendra, P.E., Epsky, N.D. (2009). Dynamics of pH modification of an acidic protein bait used for tropical fruit flies (Diptera: Tephritidae). *J. Econ. Entomol., 102,* 2371-2376.

Hogsette, J.A., Carison, D.A., Nejame, A.S. (2002). Development of granular boric acid sugar baits for house flies (Diptera: Muscidae). *J. Econ. Entomol., 95,* 1110-1112.

Hooper-Bui, L.M., Rust, M.K. (2000). Oral toxicity of abamectin, boric acid, fipronil, and hydramethylnon to laboratory colonies of Argentine ants (Hymenoptera: Formicidae). *J. Econ. Entomol., 93,* 858-864.

Hyrsl, P., Buyukguzel, E., Buyukguzel, K. (2007). The effects of boric acid-induced oxidative stress on antioxidant enzymes and survivorship in *Galleria mellonella. Arch. Insect Biochem. Physiol., 66,* 23-31.

Hyrsl, P., Büyükgüzel, E., Büyükgüzel, K. (2008). Boric acid-induced effects on protein profiles of *Galleria mellonella* hemolymph and fat body. *Acta Biol. Hung., 59,* 281-288.

Kartal, S.N. (2010). Boron-based wood preservatives and their uses, In: M.P., Chung, Ed., *Handbook on Borates: Chemistry, Production and Applications,* New York, NY: Nova Science Publishers Inc., Chap. 10.

Kaup, Y., Schmid, M., Middleton, A., Weser, U. (2003). Borate in mummification salts and bones from Pharaonic Egypt. *J. Inorg. Biochem., 94,* 214-220.

Kilani-Morakchi, S., Aribi, N., Farine, J.P., Smagghe, G., Soltani, N. (2006). Cuticular hydrocarbon profiles in *Blattella germanica*: effects of halofenozide, boric acid and benfuracarb. *Commun. Agric. Appl. Biol. Sci., 71,* 555-562.

Kistler, R.B., Helvaci, C. (1994). Boron and borates. In: D.D. Carr, Ed., *Industrial Minerals and Rocks,* 6th ed., Littleton, CO: Society for Mining, Metallurgy and Exploration, Inc., pp. 171-186.

Klotz, J.H., Greenberg, L., Amrhein, C., Rust, M.K. (2000). Toxicity and repellency of borate-sucrose water baits to Argentine ants (Hymenoptera: Formicidae). *J. Econ. Entomol., 93,* 1256-1258.

Klotz, J.H., Moss, J.I., Zhao, R., Davis, L.R. Jr., Patterson, R.S. (1994). Oral toxicity of boric acid and other boron compounds to immature cat fleas. *J. Econ. Entomol., 87,* 1534-1536.

Larcan, A., Lambert, H., Laprevote-Heully, M.C., Nida, F. (1975). Acute intoxication by cosmetics. *Eur. J. Toxicol. Environ. Hyg., 8,* 265-274.

Lizzio, E.F. (1986). A boric acid-rodenticide mixture used in the control of coexisting rodent-cockroach infestations. *Lab. Anim. Sci., 36,* 74-76.

Lloyd, J.D. (1998). International Research Group on Wood Preservation: IRG/WP/98-30178; IRG Secretariat, Stockholm, Sweden.

Lloyd, J.D., Dickinson, D.J. (1992). International Research Group on Wood Preservation: IRG/WP/1533; IRG Secretariat, Stockholm, Sweden.

Lloyd, J.D., Dickinson, D.J., Murphy, R.J. (1990). International Research Group on Wood Preservation: IRG/WP/1450; IRG Secretariat, Stockholm, Sweden.

Lyday, P.A. (1985). End uses of boron other than glass, In: J.M. Barker, S.J. Lefond, Eds., *Borates: Economic Geology and Production,* New York, NY: Society of Mining Engineers, AIME, Chap 7.

Maistrello, L., Henderson, G., Laine, R.A. (2003). Comparative effects of vetiver oil, nootkatone and disodium octaborate tetrahydrate on *Coptotermes formosanus* and its symbiotic fauna. *Pest Manag. Sci., 59,* 58-68.

Meder, R., Franich, R.A., Callaghan, P.T. (1999). [11]B magnetic resonance imaging and MAS spectroscopy of trimethylborate-treated radiata pine wood. *Solid State Nucl. Magn. Reson., 15,* 69-72.

O'Neil, M.J., Ed. (2001). *The Merck Index,* 13th ed., Rahway, NJ: Merck and Co., Electronic Version.

Obanda, D.N., Shupe, T.F., Barnes, H.M. (2008). Reducing leaching of boron-based wood preservatives-a review of research. *Bioresour. Technol., 99,* 7312-7322.

Raberg, U., Hafren, J. (2009). Gravimetric screening method for fungal decay of paper: inoculation with *Trametes versicolor. Biotechnol. Lett., 31,* 1519-1524.

Reilly, W.J. Jr. (2006). Pharmaceutical necessities, In: R. Hendrickson, Ed., *Remington: The Science and Practice of Pharmacy,* 21st ed., Philadelphia, PA: Lippincott Williams & Wilkins, pp. 1083-1084, 1089.

Rosenfelder, W.J. (1978). The industrial uses of boron chemicals. *Chem. Ind., 12,* 413-416.

Rust, M.K., Reierson, D.A., Klotz, J.H. (2004). Delayed toxicity as a critical factor in the efficacy of aqueous baits for controlling Argentine ants (Hymenoptera: Formicidae). *J. Econ. Entomol., 97*, 1017-1024.

Stanley, M.C., Robinson, W.A. (2007). Relative attractiveness of baits to *Paratrechina longicornis* (Hymenoptera: Formicidae). *J. Econ. Entomol., 100*, 509-516.

Strong, C.A., Koehler, P.G., Patterson, R.S. (1993). Oral toxicity and repellency of borates to German cockroaches (Dictyoptera: Blattellidae). *J. Econ. Entamol., 86*, 1458-1463.

Temiz, A., Alfredsen, G., Eikenes, M., Terzier, N. (2008). Decay resistance of wood treated with boric acid and tall oil derivatives. *Bioresour. Technol., 99*, 2102-2106.

Thierauf, A., Serr, A., Halter, C.C., Al-Ahmad, A., Rana, S., Weinmann, W. (2008). Influence of preservatives on the stability of ethyl glucuronide and ethyl sulphate in urine. *Forensic. Sci. Int., 182*, 41-45.

Ulloa-Chacon, P., Jaramillo, G.I. (2003). Effect of boric acid, tipronil, hydramethylnon, and difluobenzuron baits in colonies of ghost ants (Hymenoptera: Formicidae). *J. Econ. Entomol., 96*, 856-862.

Unger, A., Schniewind, A.P., Unger, W. (2001). *Conservation of Wood Artifacts*, Berlin: Springer.

United States Pharmacopeia 30 / National Formulary 25 (2007). Rockville, MD: United States Pharmacopeial Convention, Electronic Version.

Usta, M., Ustaomer, D., Kartal, S.N., Ondaral, S. (2009). Termite resistance of MDF panels treated with various boron compounds. *Int. J. Mol. Sci., 10*, 2789-2797.

Valera, M.C., Camargo, C.H., Carvalho, C.A., de Oliveira, L.D., Camargo, S.E., Rodrigues, C.M. (2009). Effectiveness of carbamide peroxide and sodium perborate in non-vital discolored teeth. *J. Appl. Oral Sci., 17*, 254-261.

Vimaladevi, M. (2005). *Textbook of Cosmetics*, New Delhi: CBS Publishers, Chaps. 2- 6.

Wang, C., Bennett, G.W. (2009). Cost and effectiveness of community-wide integrated pest management for German cockroach, cockroach allergen, and insecticide use reduction in low-income housing. *J. Econ. Entomol., 102*, 1614-1623.

Wang, Q., Li, J., Winandy, J.E. (2004). Chemical mechanism of fire retardance of boric acid on wood. *Wood Sci. Technol., 38*, 375-389.

Williams, L.H., Amburgey, T.L. (1987). Integrated protection against lyctid beetle infestations. IV. Resistance of boron-treated wood (*Virola spp.*) to insect and fungal attack. *Forest. Prod. J., 37*, 10-17.

Wilmott, J.M., Aust, D., Brockway, B.E., Kulkarni, V. (2005). The delivery systems' delivery system, In: M.R. Rosen, Ed., *Delivery System Handbook for Personal Care and Cosmetic Products*, Norwich, NY: William Andrew Publishing, Chap 21.

Woods, W.G. (1994). An introduction to boron: history, sources, uses and chemistry. *Environ. Health. Perspect., 102*, 5-11.

Xue, R.D., Barnard, D.R. (2003). Boric acid bait kills adult mosquitoes (Diptera: Culicidae). *J. Econ. Entomol., 96*, 1559-1562.

Xue, R.D., Ali, A., Kline, D.L., Barnard, D.R. (2008). Field evaluation of boric acid- and tipronil-based bait stations against adult mosquitoes. *J. Am. Mosq. Control Assoc., 24*, 415-418.

Xue, R.D., Kline, D.L., Ali, A., Barnard, D.R. (2006). Applications of boric acid baits to plant foliage for adult mosquito control. *J. Am. Mosq. Control Assoc., 22,* 497-500.

Zhang, Y.C., Perzanowski, M.S., Chew, C.L. (2005). Sub-lethal exposure of cockroaches to boric acid pesticide contributes to increased Bla g 2 excretion. *Allergy, 60*, 965-968.

Zurek, L., Gore, J.C., Stringham, S.M., Watson, D.W., Waldvogel, M.G., Schal, C. (2003). Boric acid dust as a component of an integrated cockroach management program in confined swine production. *J. Econ. Entomol.,*   *96,*   1362-1366.

INDEX

F

G

H

I

M

Q

R